AF807645

SOVIETS IN SPACE

TL789.8
.R95613
1974

PETER L. SMOLDERS

SOVIETS IN SPACE

Translated by

MARIAN POWELL

Foreword by

PATRICK MOORE

TAPLINGER PUBLISHING CO., INC.
NEW YORK

BAKER VANGUARD LIBRARY
BENTLEY COLLEGE

First published in the United States in 1974 by
TAPLINGER PUBLISHING CO., INC.
New York, New York

COPYRIGHT © 1971 KLUWER AGEMENE UITGAVEN,
HOLLAND

ENGLISH (REVISED) EDITION COPYRIGHT © 1973
LUTTERWORTH PRESS

All rights reserved
Printed in Great Britain

No part of this book may be reproduced or transmitted in any form or by any means, electronic or mechanical, including photocopy, recording, or any information storage and retrieval system now known or to be invented, without permission in writing from the publisher except by a reviewer who wishes to quote brief passages in connection with a review written for inclusion in a magazine, newspaper or broadcast.

Library of Congress Catalog Card Number 73-16177

ISBN 0-8008-7340-8

DEDICATION

To the memory of
Cosmonauts Georgi Dobrovolski,
Vladislav Volkov and Viktor Patsayev,
first dwellers in space.

CONTENTS

COLOUR PLATES

MONOCHROME PLATES

PREFACE

On June 7, 1971 three Russians moved into the first space dwelling. For nearly 23 days and nights the cosmonauts Georgi Dobrovolski, Vladislav Volkov and Viktor Patsayev occupied the world's first space station, Salyut. They were engaged in meteorological observations, astronomical activities, medical tests and physical exercises.

They ate, slept, listened to the radio and read their favourite books. They lived in more or less domestic conditions. Dobrovolski, Volkov and Patsayev died during that part of their flight which was gradually being regarded as a routine operation: the return to earth. The tragic death of these three pioneers, however, hardly detracts from the value of their flight which—almost imperceptibly—marked a new phase in spaceflight: the period of living in space. For this flight involved essentially new aims. Up to this time the primary purpose of spaceflights had been to test new technical apparatus, or—as was obviously the case in the American flights—to reach other planets. The latter formed part of what might by now be called classical space philosophy. Its objectives were the moon, the planet Mars, etc. Space itself was an obstacle to be overcome. None of the classical writers of space fiction—of whom H. G. Wells was the last great example—makes any mention of using space itself for any scientific or economic purpose whatsoever. But we are gradually coming to realise that there is so much space that it simply cries out to be exploited.

One of its most obvious uses is the observation of the earth. Thus we see that, paradoxically, the purpose of spaceflights is more and more earth itself. Salyut clearly confirms this new approach. In the next few years more and more Russians and Americans will "settle" in space, if only for relatively short periods. This will greatly increase the significance of manned spaceflights for the purposes of science, technology and economy. Moreover, there will be new opportunities for co-operation between Russian and American "space dwellers". This does not mean that flights to other planets will cease, but they will form part of great, harmoniously-conceived plans for using and reconnoitring interplanetary space.

The Russians in particular have always declared that they wanted to get a "firm foothold" in space before setting foot on the moon and the other planets. This book attempts to explain the motives which led to this point of view. The history of Russian spaceflight described in this book makes it clear why the Russian approach to manned spaceflights has so far deviated considerably from that of the Americans. Unmanned space exploration too is treated in detail, since it is closely linked to manned flights in the cosmos.

It is generally thought that in the Soviet Union it is impossible to obtain worthwhile information concerning spaceflights. This is only partly the case. It is true that an overall view of Russian spaceflights is not as readily available as that of American operations, but if every detail—however insignificant in itself—is recorded from year to year, subsequently to be carefully considered in relation to others, a clear view may undoubtedly be obtained. This applies not only to the text of this book, but also to the illustrations. The extremely accurate drawings supplied by Coen Benraad, Rob Koch and myself are intended to be not only informative, but above all reliable.

PETER SMOLDERS

FOREWORD

by

Patrick Moore

ON OCTOBER THE FOURTH, 1957, the Space Age began. It was started by the ascent of Sputnik-1, the Russian artificial satellite which sped round the world sending back the well-remembered "Bleep! bleep!" signals which were picked up in every country with modern-type technology. Progress since then has been amazingly rapid; within a dozen years after that first Sputnik, men had landed on the moon and rocket probes had been sent out to the planets.

To all intents and purposes, there have been only two "nations in space": the USSR and the USA. Other countries have sent up small satellites, but virtually all the progress has been either Russian or American. The two programmes have been developed quite independently, mainly because of political complications, and each has had its quota of remarkable successes, together with inevitable failures and a few tragedies. Moreover, it is notable that the programmes have not been working along the same lines. No Russian has yet tried to reach the moon, while no American probe has made a parachute descent through the dense, cloudy atmosphere of the planet Venus.

In those early days there was much talk of a "space race", and the idea still lingers on, though there is little justification for it. Of course, publicity has been different. The Americans have always given details of their plans well in advance, and their lunar missions have been shown "live" on television. The Russians are much less forthcoming, and at least part of their space programme has been somewhat shrouded in mystery, particularly as the usual practice is to give details after the event instead of before.

The result of this is that although almost countless books have been written about American astronautics, relatively little has come out of the Soviet Union. This is one reason why I welcome the present book. The author is Dutch, but he has had close contact with the Russians, and he writes with authority. Also, he is pleasingly impartial; he is not afraid to criticise, but he has rightly refrained from seeing "space politics" where they do not exist.

This does not set out to be a detailed technical handbook, but as a general introduction to the story of Soviet space research during this all-important pioneer period it succeeds admirably, and it will be extremely useful to a wide variety of readers.

What of the future? At present (1973) Russia and America are still the only nations with a really serious space interest. This will be altered in the future, but already there are encouraging signs of co-operation, and a joint Soviet–US space venture has already been officially announced. In a way, then, we have reached the end of one era and are about to enter another. The story of Russian achievements, as given here, is part of history—and the author has told it well.

PATRICK MOORE

It is of course difficult to imagine future space travel without manned spacecraft. Space cannot possibly be explored by automatic luniks and interplanetary probes only. They have undoubtedly brought masses of new facts to light and will tell us a great deal more in the future, but in my opinion they are merely the first step in the reconnaissance of space. Only man is able to ensure the complete success of this enterprise.

YURI GAGARIN

SPACE TRAVEL: THE SOVIET APPROACH

The first space dwellers – Space Hero Number One – Rumours – Organisation – Efficiency and Standardisation – Space Station : the focal point of the programme – Russians to the moon

The capsule had made a soft landing, but the three men inside lay motionless in their contoured seats. The small lamp over the instrument panel threw a faint light on their white faces. It was thus a tragic sight which met the doctors of the recovery team in the steppes at dawn.

Georgi Dobrovolski (43), Vladislav Volkov (35) and Viktor Patsayev (38) had paid with their lives for their amazing, record-breaking flight on board the world's first space station, Salyut.* An "unfair" price to pay, as the poet Yevgeniy Yevtushenko was to say. Georgi, Vladislav and Viktor had become the first space dwellers, the first men from earth to settle temporarily in space, in order to study, from their cosmic home, their mother planet and the universe.

Nothing in the course of their cosmic marathon gave any indication of the dramatic way in which their journey was to end. The unceasing stream of medical information which reached the earth showed that the men stood up well to the effects of lasting weightlessness. They proved that it is possible to live in space for a considerable length of time; that space can indeed be used for the purpose of observing the earth, its atmosphere and its natural resources; that space is not merely an obstacle to be overcome in order to reach the moon and the planets, but that it may be made subservient to science, economics, meteorology, astronomy, geology, biology. This fact introduces a new era in manned space exploration, an era in which, paradoxically, it will for the time being be earth itself, and not another planet, which will be the main objective of travellers in space.

The flight of Dobrovolski, Volkov and Patsayev lasted more

*See: "A Firm Foothold in Space", p. 238.

than 570 hours; 569 hours of this passed without essential problems. Death struck in the last hour of their flight. The braking rockets had already done their work when suddenly the air escaped from the three-man Soyuz-11 capsule. Within a few seconds the oasis of life, which the atmosphere in that small capsule had provided, was dissolved in the vast desert of space. Death must have struck so suddenly that Georgi, Vladislav and Viktor can hardly have realised what was happening.

The next day their bodies lay in state in Moscow. Foreigners as well as thousands of Russians rendered them the last honours. All mankind sympathised with the cosmonauts' next-of-kin: with their parents, with Ludmilla Dobrovolski, her daughters Maria (12) and Natasha (4), with Ludmilla Volkov and her 13-year-old son Vladimir, with Vera Patsayev and her children Dimitri (14) and Svetlana (9).

It came as a shock to realise that manned spaceflights, which many people had gradually come to regard as routine operations, were still risky undertakings. Lack of detailed information at first gave rise to the belief that the Soyuz-11 cosmonauts had been unable to withstand the deceleration forces during their re-entry into the atmosphere. Dobrovolski, Volkov and Patsayev had, after all, lived in weightless conditions longer than anybody else. But it soon became clear that their death was due to technical and not to physical causes. This was an important and positive conclusion. It meant that the exploit of these three brave Russians had not been in vain. The Russians would continue their exploration of space. Even before the cosmonauts had been taken to their last resting place in the wall of the Kremlin,* Pravda wrote: "We know that after this painful loss, the difficult and dangerous fight against nature will be carried on with undiminished determination and consistency. The Soviet people are accustomed to struggle and do not shrink from obstacles."

By the end of September 1973 the Soviets proved by the two-day flight of Soyuz-12 that they had overcome the difficulties.

*The funeral was attended, among others, by the American astronaut Tom Stafford.

1. (*Above*) The crew of the first space station, Salyut. From l. to r.: Viktor Ivanovich Patsayev, Georgi Timofeyevich Dobrovolsky and Vladislav Nikolayevich Volkov. They died in their spaceship Soyuz-11 during the return to earth.

(*Below*) Artist's impression of the link-up between the spaceship Soyuz-11 and the space station Salyut (in the foreground).

2. (*Above*) *Left* Sergey Pavlovich Korolyov, chief designer of Soviet rockets and spaceships. He died in 1966, aged 60. *Right* Mstislav Keldysh, president of the Academy of Science of the USSR and responsible for the Soviet space programme.

(*Below*) The centre of Zyvezdograd (Star City), a town of scientists and technicians. The building of this town began in the mid-fifties.

3. The launch centre at Baikonur. A Soyuz rocket (Soyuz-10) being transported to the launching pad.

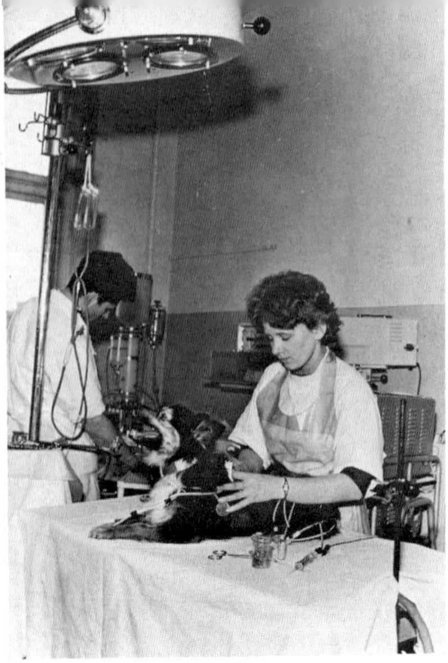

4. (*Above*) *Left* Medical examination of the dog Ugolyok before its flight in Cosmos-110. *Right* Konstantin Eduardovich Tsiolkovsky (1857–1935), the father of space travel.

(*Below*) A standard Cosmos satellite, complete with solar cell panels. Considerable standardisation has been achieved in the programme of light scientific Cosmos satellites.

Space Hero Number One

One year before the accident. It is May and in Moscow it has been a warm day. Hundreds of Moscovites are sauntering along the Prospekt Mira, enjoying the fresh evening air. A slender rocket, towering high on a hundred metre stainless steel column, points into space. The Monument for the Vanquishers of the Cosmos is bathed in the light of enormous floodlights.

The entrance to the path leading to this dazzling sculpture is flanked by the statues of those who, in the name of the Soviet Union, conquered the cosmos: Yuri Gagarin, Valentina Tereshkova, Pavel Belyayev, Alexei Leonov, Vladimir Komarov. . . . Three of these five pioneers are no longer alive. Gagarin died on March 27, 1968 during a training flight in a Mig-15. Belyayev died of a stomach disease and Komarov crashed on April 24, 1967 in the first prototype of the Soyuz series, which later proved to be so successful.

A young couple buy a small bunch of flowers from the flower-stall opposite the Underground Station, and place it in front of the statue of their countryman who, on April 12, 1961 was the first man to penetrate into the universe. A simple gesture, but more significant than the official laying of a wreath. To the Russians, Yuri is still Space Hero Number One. Just over 12 years ago he became the first man in space. His flight will remain a unique achievement, of its kind never to be surpassed, just as Neil Armstrong's first step on the moon can never be surpassed.

A great deal has happened in these 12 years. Gagarin's successors made flight after impressive flight into space, at first hardly bothered by American competition. However, in 1965 the United States began to launch their Geminis at an incredible rate. President Kennedy had accepted the Russian "challenge" and America was making double-quick time to the moon. In retrospect it is doubtful if their urgency was justified, for the Russians were not in all that much of a hurry and—what is more important—they were on an entirely different tack. . . .

But in the middle sixties this was obvious neither to the Americans, nor to anybody else in the West, nor even to most people in the Soviet Union. In any case the Russians appeared to be very self-confident and to feel themselves superior to the

Americans. From time to time, moreover, they casually mentioned the moon and that in itself was sufficient reason for the top NASA people across the Atlantic to increase their efforts.

Rumours

The Russians were not very forthcoming about the background to their space programme. They did not publish any timetable, they never announced their flights in advance and they never once made their exact objectives clear. And this was particularly so in the early stages of what the Americans persisted in calling the "race to the moon".

Since then the Soviets have become somewhat more talkative. Although complete silence in advance produced the maximum element of surprise after each new spaceflight, it clearly had its disadvantages. At first the public got the impression that the Soviets could achieve anything they wanted in the field of space exploration. This led to the wildest rumours concerning the Russians' plans. If the results of the latest flight did not come up to these expectations, they created a suspicion that the flight had been a failure.

This kind of speculation, though of course entirely unjustified, is nevertheless understandable. If the Americans had not announced before every Apollo flight what they intended to do, an operation like that of Apollo-8, the first flight round the moon, would undoubtedly have given rise to rumours that a landing on the moon had failed. Apollo-9—the first test of command module and lunar capsule close to earth—would, after Apollo-8, have given rise to the rumour that this time the Americans had not even succeeded in getting out of earth orbit. And Apollo-10 (approach to within 15 km of the moon's surface by the LM) would certainly have led "expert observers" to conclude that America was having great difficulties with its moon programme; the landing would have been said to have failed "at the last moment".

This is the situation in which the Soviets used to find themselves to some extent, although lately they are often a little more forthcoming about their intentions. Thus they announced at the time of launching Soyuz-9 that the operation was to be a solo flight in earth orbit, thereby scotching in advance all speculation concerning future launchings.

The worst rumours which used to go the rounds again and again concerned cosmonauts who were said to have perished. None of these speculations were based on the slightest evidence. Usually they concerned flights which had not been announced by the Soviet Union even after launching, as if it had been known in advance that the operation would fail. Occasionally the rumour concerned a known launching, such as that of Sputnik Spacecraft-1 on May 15, 1960. This was the first test flight of a spaceship of the Vostok type. Rumour had it that there were no fewer than three (!) people on board. After a flight of four (!) days, the braking rockets were fired, but in the wrong direction. As a result the spaceship went into a higher orbit and did not return to earth. The latter point was quite true, but there was nothing to indicate that there were human beings on board. Moreover, why should the Russians not have announced this? In any case it was not until almost a year later, after four more unmanned Vostok flights, that one man, Yuri Gagarin, was to make a solo flight in a Vostok lasting exactly one and a half hours. Before the end of this flight Vostok's launching was known of all over the world. The Soviets subsequently adhered to this line of conduct, and when Vladimir Komarov became the first victim of space, this too was announced by Tass. Of course the Russians, like the Americans, met with difficulties in their conquest of space. At times there were problems in the cosmonauts' training. Cosmonaut Leonov, for instance, told me in Moscow about a parachute jump which had nearly cost him his life. Leonov said: "It happened while we were preparing for the Vostok flights. We had to train with the ejector seat since cosmonauts had the opportunity to leave the spaceship during its descent. In order to be able to use our parachutes, we had to be catapulted from our contoured metal seats for the landing. While jumping, one of my straps caught on the seat, so that I would have landed on my head or my stomach. I realised that I would not come off too well. I therefore tried to bend the back of the seat, which was made of duralumin. I managed to free the strap at a height of 45 m and made a perfect landing. Later on it needed a hammer to bend the metal back!"

Testing unmanned craft did not always go smoothly either. There were problems with earth satellites, and a number of lunar- and planetary-probes were lost in the interplanetary waste. Some of these were obvious cases, others less so. It is

Fig. 1. Space centres in the Soviet Union.

Star village lies approximately 70 km N.E. of the centre of Moscow. This is where the cosmonauts prepare for their flights.
Star City (Zvyezdograd) is situated near the launching base of Baikonur. Here live the scientists and technicians involved in the launching and control of manned and unmanned spacecraft.
Plesetsk, 170 km south of Archangel, is the launching base for lightweight scientific satellites and artificial moons for research purposes which travel in more or less polar orbits.
Kapustin Yar, in the steppes East of Stalingrad, is used mainly for launching lightweight scientific satellites.
Baikonur, the chief launching base. This is where all manned spaceships are launched, as well as lightweight and heavy unmanned spacecraft.

therefore pointless to enumerate them all; such an enumeration would in any casc be extremely uninteresting. Failures—and particularly failures of unmanned craft—are of interest only in as far as they contribute to the success of a subsequent experiment.

Organisation

In the past few years in particular, the organisation and structure of the Soviet space programme has become increasingly clear, at least to those who have taken the trouble to study it. Like most activities in the Soviet Union, spaceflight is controlled by the Central Committee, which passes its directives via the Supreme Soviet to the Council of Ministers. This Council is assisted by, among other bodies, GOSPLAN, the state planning bureau responsible for planning scientific and economic activities as well as for budget control.

The co-ordination of all space activities, manned as well as unmanned flights, is in the hands of the Soviet Academy of Sciences. The Academy is directed by Professor Mstislav Keldysh (62). The Academy advises the Council of Ministers on planning and budgeting. As far as space travel is concerned, contact with government bodies takes place through the State Commission for Science and Technology (Chairman A. Kirilin) and the Interdepartmental Commission for Interplanetary Communications, of which Leonid Sedov was for a long time the Chairman. The latter is now Chairman of the Commission for Space Travel, a special commission of the Academy. There are a few more such special commissions: the Commission for Peaceful Space Exploration, Chairman Professor Anatoly Blagonravov, and the Commission for International Co-operation in Space, of which Professor Boris Nikolayevich Petrov is Chairman. All these men are *Akadyemiki*. The Academy has 565 Full and Associate members. Apart from this main Academy, there are academies in each of the fifteen republics of the Soviet Union. The Academy of Sciences of the USSR controls, directly or indirectly (via the academies of the various republics), 4700 scientific institutions. In all there are approximately a million scientists with varying qualifications in the Soviet Union. Under the direction of the

Academy, both theoretical and applied research is carried out. Numerous institutes contribute indirectly to the development of space travel, for instance in the fields of biology, computer science, mechanics and astronomy. The Academy also manages the twelve tracking stations within the Soviet Union, the Centre for Long-Range Space Communications in the Crimea (for flights to the moon and beyond) and seven tracking ships equipped with the most up-to-date electronic apparatus. These ships enable the Soviet Union to keep track of its space craft if necessary 24 hours a day. Unlike the Americans, the Russians do not have a chain of permanent tracking stations around the earth. The Co-ordination and Computing Centre, which processes all information received from space, also comes under the Academy. The same applies to the cosmonaut village of Zvyezdni Gorodok (Star Village), with its 3100 inhabitants, 70 kilometres from the centre of Moscow. This is where the cosmonauts—about sixty in all—live and where they are trained for their spaceflights. For ten years General Nikolai P. Kamanin was in command of the team of cosmonauts. In 1971 his position was taken over by Cosmonaut Vladimir Shatalov.

The Academy maintains contacts with institutes, design bureaus and factories which come under the various industrial ministries. In some of these factories the manned and unmanned spacecraft and the launching rockets are built. Apart from the Academy, both the army and the air force are involved in the space industry in an executive capacity. The strategic rocket forces of the army carry out the launchings and the air force assists in training the cosmonauts and in the recovery of manned and unmanned spacecraft. All these activities are carried out in consultation with the Academy. Apart from numerous small launching pads for (military) rockets, the Soviets have three large launching bases: Baikonur, Kapustin Yar and Plesetsk. Baikonur was used for the first time in 1957, when the first intercontinental missile was fired, to be followed soon after by the launching of Sputnik-1. Baikonur (longitude 65 degrees, latitude 48 degrees) lies 250 km north-east of the Aral Sea. The launching base, which westerners sometimes call "Tyura-Tam", is commanded by General Kuznetsov. This is where all manned spacecraft, all moon- and planet-probes and all heavy unmanned earth satellites are launched. Occasionally light scientific artificial satellites and

5. (*Above*) Weather satellite of the Meteor type, equipped with large, hinged solar cell panels and an umbrella-shaped aerial.

(*Below*) The heaviest unmanned Russian satellite: the 17-ton Proton-4. This artificial satellite and others in the Proton series were used for studying primary cosmic radiation.

6. (*Above*) Yuri Alexeyevich Gagarin (1934–1968). On April 12, 1961 he orbited the earth in Vostok-1. He died on March 27, 1968 during a training flight in a Mig-15.

(*Below*) Assembly-line of the Vostok spacecraft. In the foreground: finishing touches are added to two capsules. In the background: assembly of the instrument section.

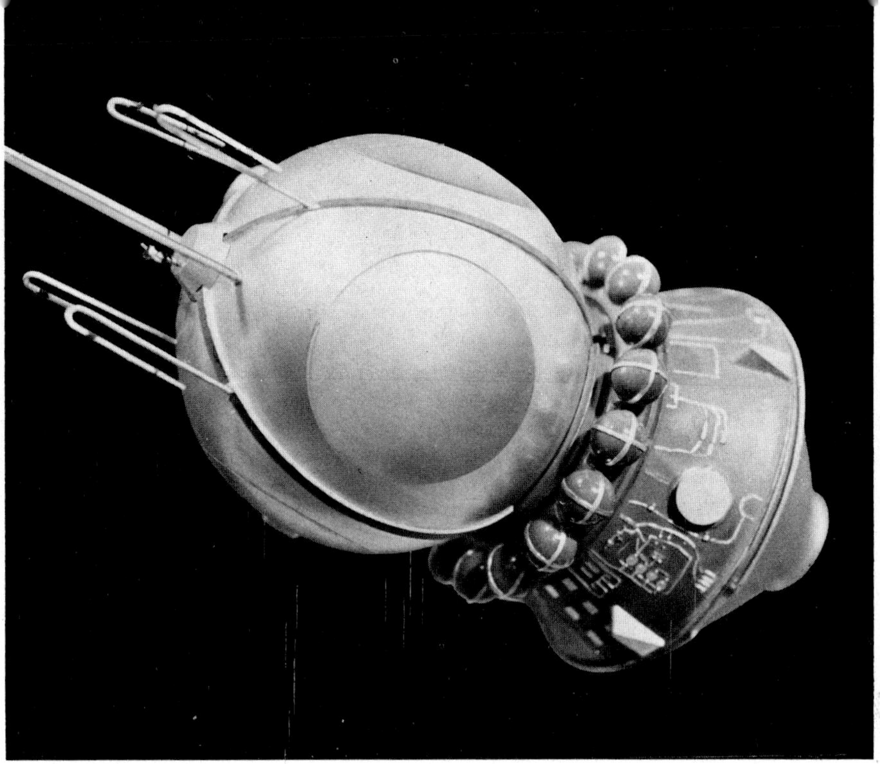

7. (*Above*) Model of a Vostok spaceship. *Left* the spherical cabin with antennae. *Right* the service compartment (instrument section), containing power cells, fuel tanks, steering rockets and a retro-rocket. The spherical bottles contain, among other things nitrogen for the altitude control system.

(*Below*) Ejector seat, as used in Vostok spaceships.

8. (*Above*) Valentina Tereshkova and her flying instructor Vladimir Sergeiyevich Seryogin, who later crashed with Gagarin in a Mig-15.

(*Below*) The crew of Voskhod-1. Nearest the camera: Vladimir Komarov. Behind him Konstantin Feokistov and Boris Yegorov.

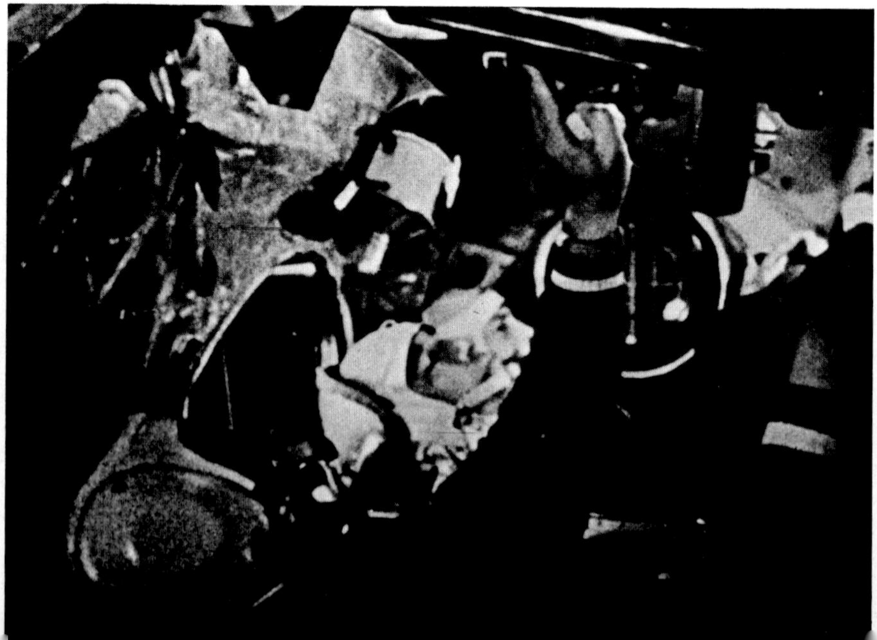

military rockets for experimental purposes are sent up from here. Thc scientists and technicians who run the base live in nearby Zvyezdograd (Star City). The building of Zvyezdograd started about 15 years ago. By now it is an average-sized town with large industrial complexes, a cinema, various grades of schools, a television- and radio-studio, and, like practically all Russian towns, a large central square. Zvyezdograd consists mainly of white buildings and when approached from a distance it appears to rise like a *fata morgana* from the surrounding steppe.

Kapustin Yar is the oldest launching base in Russia. It lies 200 km east of Volgograd, north of the Caspian Sea. In the early fifties the first meteorological rockets and animal-carrying experimental rockets were launched from Kapustin Yar for short "space jumps". Now, lightweight satellites in the Kosmos-series start from here.

The Plesetsk base is the newest of the three. This complex is situated near the White Sea at latitude 63 degrees, longitude 41 degrees, 170 km from Archangel. The base was constructed for the launching of satellites which are to travel at a large angle to the equator. Baikonur and Kapustin Yar are too far south for this kind of operation. Plesetsk is used particularly for the launching of artificial satellites for meteorological purposes and for reconnaissance satellites, which travel in approximately polar orbits. As the earth revolves within this orbit, each base passes underneath these photographic robots twice a day.

Efficiency and Standardisation

Although the centralised nature of the Russian spaceflight organisation excludes the kind of healthy scientific and technical competition which exists among American space industries, the system undoubtedly has its advantages, particularly as far as efficiency and standardisation are concerned. In the United States various space industries receive large grants for research projects, but on completion of the research only one of the companies concerned gets an order. This sort of thing is unknown in the Soviet Union. The Soviet system therefore fosters efficiency and continuity in the execution of the programme.

Fig. 2 and 3. Soyuz launch centre at Baikonur.

The rocket is conveyed on a transporter (14) and is lowered into a launching pit, so that the lower part is hidden. The rocket rests on four supports constructed on the low support platforms (9) which, like the other servicing towers, are mounted on a large revolving ring round the pit. The rocket and its surrounding installation can thus revolve horizontally (in *azimuth*).

Key to diagram:

1. Concrete platform.
2. Base of swivelling mounting frame (9).
3. Fuel trucks.
4. Concrete base of launching platform.
5. Exhaust deflector.
11. Separately rotating upper section of 10.
12. Stabilisers, used in an emergency if the spaceship is pulled clear of the carrier rocket by the escape rockets mounted above.

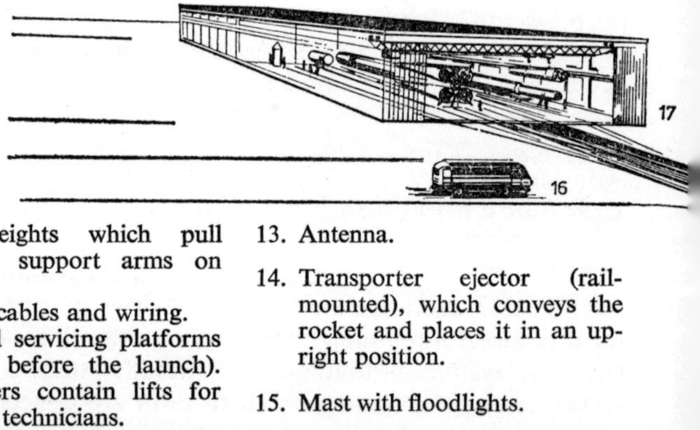

6. Counterweights which pull back the support arms on lift-off.
7. Tube for cables and wiring.
8. High-level servicing platforms (removed before the launch). The towers contain lifts for crew and technicians.
9. Swivelling low-level mounting structure.
10. Swivelling tower for fuel supply and electrical connections (this is swung away before launching on the command "Zemlya bort!")
13. Antenna.
14. Transporter ejector (rail-mounted), which conveys the rocket and places it in an upright position.
15. Mast with floodlights.
16. Cosmonauts' bus.
17. Assembly building where rocket and spaceship are assembled in a horizontal position. The building is linked to the launching pad by rails.

This is even more the case as regards standardisation which in the Russian programme is carried to great lengths, resulting in enormous financial savings. This is possible as there is effectively only one complex for the space travel industry. The standardisation is very obvious in the design of the RNV, the Vostok carrier rocket which was and is used in many different versions for launching unmanned earth satellites, heavy satellites of the Kosmos type, communication satellites, moon probes, Vostoks, Voskhods, Soyuz craft and planet scouts. In the United States a specially designed rocket is used for practically every category of spacecraft, and all these rockets are built by different companies, as are their loads.

Standardisation is also applied to a great extent in the light Kosmos satellites. These are based on a number of main components which can be used in various combinations. All this

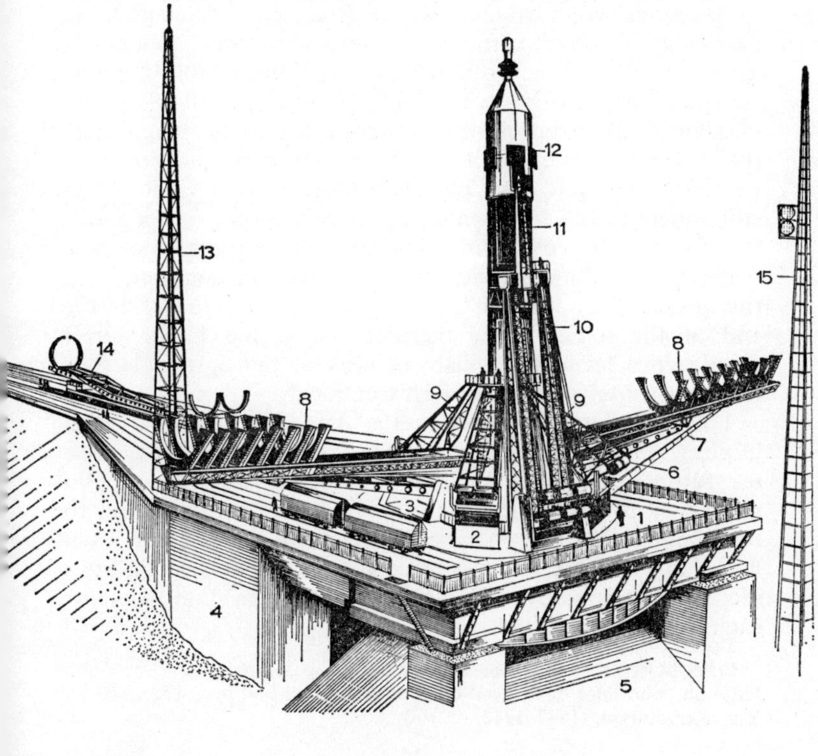

results in economic production and increased reliability. General Kamanin told me about the Russian spaceship Soyuz: "Several variants are possible. Soyuz can be used for different purposes. This spacecraft is manufactured in series and is therefore very economical." The same applies to the standard launching rocket which has been used on hundreds of occasions.

Space Station: the focal point of the programme

In the world of space travel names like Tsiolkovsky, Oberth and von Braun are well known. These men not only did important work in the field of rocketry, but they also occupied themselves intensively with the general lines on which space travel was to develop.

Although von Braun was undoubtedly influenced, in particular by Oberth, these three pioneers came, practically independently of each other, to the conclusion that the first step in manned space travel would have to be the establishment of a large space station in orbit round the earth. From there the universe could be further reconnoitred and journeys to the moon and the planets undertaken. The Russians still adhere to this theory and the space station concept forms the focal point of their manned space programme. The Americans on the other hand have (for the time being) skipped this phase. They wanted to set foot on the moon before the end of the sixties. They therefore chose the Lunar Orbit Rendezvous technique which, in view of the time-table they had set themselves, was the best solution.*

The technique developed in the Apollo programme is an impressive one and undoubtedly provides significant experience for future, even greater, space operations. But the Apollo procedure is, in essence, far too expensive and incidental to be maintained. An enormous rocket is needed for each launching and the space capsule cannot be used again after recovery. It is rather like using a brand-new jumbo-jet for each transatlantic flight.

*The LOR technique was suggested in the United States by John Houbolt, who later discovered that the same idea had been proposed by Yuri Kondratyuk (1887–1942).

The Americans fully realised the limitations imposed by their Apollo programme. They reduced the number of Apollo flights to seventeen and are now going to devote their—considerably reduced—budget to the development of that with which they should have logically started: the space station.

The Russian manned space programme, as carried out so far, has progressed logically towards the construction of permanent bases in space, of which Salyut is the precursor.

The *Vostok* flights proved that it is possible to live in space for a considerable length of time. Moreover in the framework of this project the first necessary tests were made for the realisation of a rendezvous, an essential element in the establishment of space stations.

Voskhod formed an intermediate phase. The versatile Soyuz was not yet ready and a modified Vostok—for that is what the Voskhod amounted to—was used for testing systems for a multi-crew capsule and for a sensational experiment: the first walk in space. In the construction of space stations cosmonauts would have to venture outside the safe shelter of their capsule.

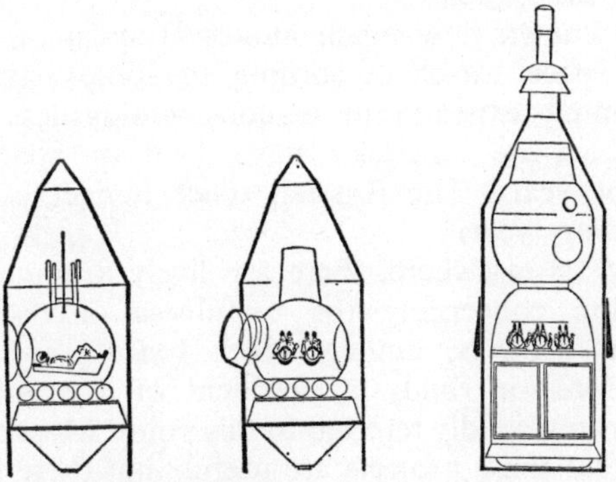

Fig. 4. Vostok, Voskhod and Soyuz.

Three manned Russian space capsules, each drawn to the same scale. The protective nose-cone, which is jettisoned outside the earth's atmosphere is also shown. Voskhod-2 is typical of the Voskhod series. It is shown together with its air-lock retracted and covered.

Soyuz made it possible to carry out extensive manoeuvring-, rendezvous- and link-up operations, in preparation for the construction of large stations in space. The cosmonauts also trained in stepping from one spaceship to another and welding experiments were carried out, which would be necessary in the assembly of permanent space ports.

In the field of unmanned spaceflights, too, the Russians have developed an extensive programme, which is being executed with ever-increasing speed. American space research, on the other hand, since 1966 is becoming less intensive. In 1970 81 launchings took place in the USSR as against only 29 in the United States.

Apart from purely scientific experiments the Soviets pay a great deal of attention to space projects with immediate practical benefits, such as weather- and communication satellites. The flights of Luna-16, Luna-17 (Lunokhod-1) and Venera-7 have clearly shown of what the Russians are capable in the field of unmanned reconnaissance.

In this area the Soviets have opened up immense possibilities, which promise great things for the near future. Objectively considered, it must be admitted that the Russian space programme as a whole appears to be carefully balanced. Constant expansion is anticipated and the various phases follow each other in logical sequence.

It is not known how much money is spent in the Soviet Union on space travel. According to some estimates, the average annual expenditure is considerably less than the amounts which the Americans have spent on spaceflights in the past few years. The Russian space budget is, however, increasing every year.

Naturally, as elsewhere, there are lively arguments in the Soviet Union concerning the usefulness of research and development projects, among which space travel must be counted. Professor Keldysh, president of the Academy of Sciences, has repeatedly referred to this point. Most experts are convinced that these projects are useful, but there is no unanimity on how investments in research and development may be compared with results. In connection with the 1971–75 five-year plan, Vadim A. Trapeznikov, vice-chairman of the State Commission for Science and Technology, has calculated that each rouble invested in research and development would result in an annual increase of the national income of 1·49 roubles, as against only 0·39 roubles if the money is invested directly in the economy. This is a scientist's view. Some of the officials of GOSPLAN are less optimistic. They say that in 1966 each rouble invested in research and development was expected to yield a 0·74 per cent profit, but that it proved to be less in reality. Inevitably scientists and space experts will get less than they would like, but that is after all not so very surprising.

Russians to the moon

When the American astronauts Neil Armstrong and Edwin Aldrin became the first men to set foot on the moon, some people took the view that the Russians had at the last moment "wisely" withdrawn from the so-called "race for the moon". If the publications and statements of Russian experts in the last ten years are carefully analysed, it becomes clear that there are no grounds for this view.

It is true that shortly after Gagarin's flight the Russians made no secret of their conviction that the first man on the moon would be a Russian. They were not the only people to think so—the Americans did too. At that time Wernher von Braun made his famous pronouncement: "I am convinced that when we arrive on the moon we shall have to pass through Russian Customs". At that moment it was of course very difficult to anticipate that the Americans would be the first. While the Russians amazed the world with a series of impressive spaceflights, the Americans initially confined themselves to following their rivals at a distance. Already in the early sixties the Russians had announced that their moon-rocket would be assembled in earth orbit and that they regarded the establishment of a space station as the essential initial phase in the reconnaissance of space. They had therefore opted for what in the United States was called "Earth-Orbit-Rendezvous". The Americans subsequently chose the "Lunar-Orbit-Rendezvous" technique. The reason for both procedures was that a direct flight—i.e. without an intermediary stage—was, from a practical point of view, virtually impossible.

America was worried.

Exactly eight days after Gagarin had been the first human

being to circle the earth in cosmic orbit, President Kennedy sent Vice-President Lyndon Johnson, who was then Chairman of the Space Council, the following memorandum:

MEMORANDUM FOR VICE-PRESIDENT

In accordance with our conversation I would like for you as Chairman of the Space Council to be in charge of making an overall survey of where we stand in space.

1. Do we have a chance of beating the Soviets by putting a laboratory in space, or by a trip round the moon, or by a rocket to land on the moon, or by a rocket to go to the moon and back with a man. Is there any other space program which promises dramatic results in which we could win?
2. How much additional would it cost?
3. Are we working 24 hours a day on existing programs. If not, why not? If not, will you make recommendations to me as to how work can be speeded up.
4. In building large boosters should we put our emphasis on nuclear, chemical or liquid fuel, or a combination of these three?
5. Are we making maximum effort? Are we achieving necessary results?

I have asked Jim Webb, Dr. Wiesner, Secretary McNamara and other responsible officials to co-operate with you fully. I would appreciate a report on this at the earliest possible moment.

JOHN F. KENNEDY.

On May 25, 1961 Kennedy announced that the United States would make every effort to "land a man on the moon and bring him safely back to earth before the end of the decade". In the years that followed, the Americans showed that they were in earnest. Enormous factories and gigantic launching bases were built; hundreds of thousands of people earned their living in the rapidly expanding space industry. And in spite of setbacks, two Americans landed on the moon well within the time limit they had set themselves.

The Americans had won the much-discussed space race. But . . . can it be said that there was more than *one* participant in the race? Sadly, we must doubt it. Russian expectations, initially published with so much optimism, that they would be the first to set foot on the moon, were based solely on the situation as it was at that moment, when nobody thought that the Americans had much chance. But once it became obvious that Kennedy was in earnest, the Soviets soon made it clear that they had no intention of being pressured by the tempo of the Americans. As early as 1963 Khrushchev declared openly that the Russians would be in no hurry to get to the moon and that they had no intention of competing with the Americans in this field. On June 17 of that year Sir Bernard Lovell, Director of the large radio-telescope at Jodrell Bank, returned from a study tour in the USSR, in the course of which he had visited the Centre for Remote Space Communications in the Crimea. At a press conference he declared emphatically that for the time being Soviet scientists had opted for unmanned moon reconnaissance. On October 7 of the same year Professor Sedov said in Brussels that his comrades had no immediate intention of going to the moon. In any case, he said, no special moon-landing system had yet been developed. On October 13, 1966, at a congress of the International Federation of Astronauts in Madrid, he again declared that the Soviet Union did not intend to compete with the United States as far as a manned flight to the moon was concerned. He denied emphatically that the Russians planned to put a man on the moon before 1970. On the other hand, Sedov stated that he thought a manned American moon-landing before that year was a possibility. "It is of no consequence who arrives first, we or the Americans," he said. "I am waiting in suspense for details of the American plans." Meanwhile the Soviets have energetically embarked on unmanned moon reconnaissance, and there is no doubt that in this field their efforts will outlast those of the Americans, who, now the remaining Apollo flights have been completed, have no further plans for either manned or unmanned reconnaissance of our neighbour in the universe.

It is said that the Russians give preference to unmanned moon research, but this may be doubted. Certainly they are not considering landing men on the moon just yet, but at the same time there is no doubt that they will send cosmonauts to the moon as soon as they have established suitable space stations as a starting point. First, and above all, they wish to have a firm foothold in space. When I asked space-walker Alexei Leonov whether a permanent space station would at some time in the future serve as a starting point for manned flights to the moon and the planets, he smiled in amazement and exclaimed: "Konyechno! (of course!). To the moon, to the planets, where you will!"

The Russians are truly of the opinion that with currently available means manned moon research can only lead to very expensive experiments on a modest scale. Once a fully-fledged space-transport system has been evolved, based on space stations and repeatedly usable shuttle-rockets, cosmonauts will be able to explore the moon fully, at a fraction of the cost of present-day manned flights. The Russians are right in saying that the possibilities of reconnaissance by robots have not nearly been exhausted. On the other hand it is obvious that they have been placing more emphasis on the advantage of unmanned reconnaissance since the success of America's Apollo astronauts. In any case this account is not meant to suggest that the Russians have taken no notice of the American space programme. There are several indications that on many points of detail they certainly behaved like rivals; consider for instance the unsuccessful mission of Luna-15 at the time of Armstrong's first steps on the moon. Nor can it be pure coincidence that the Russians have never carried out an important experiment which had previously been made by the Americans. If at the time Beregovoi had linked up his Soyuz-3 to the unmanned Soyuz-2 capsule, it would have given rise to the remark that NASA had long since carried out a similar operation within the framework of the Gemini-project. Consequently the Russians waited with their link-up in space until the flight of Soyuz 4–5, thus being the first to link up two manned spaceships.

In all this it was to the Soviets' advantage that the Americans usually announce their plans well in advance. An important exception to this rule was the Apollo-8 mission, the first flight round the moon at Christmas 1968. At that time there were plenty of indications that the Russians were planning a similar flight, which requires far less effort than a moon landing. Towards the end of 1968 Dr. Charles Sheldon, America's most distinguished expert on Russian spaceflight, had made an official report to the US House of Representatives. He expressed the conviction that the Russians would attempt a

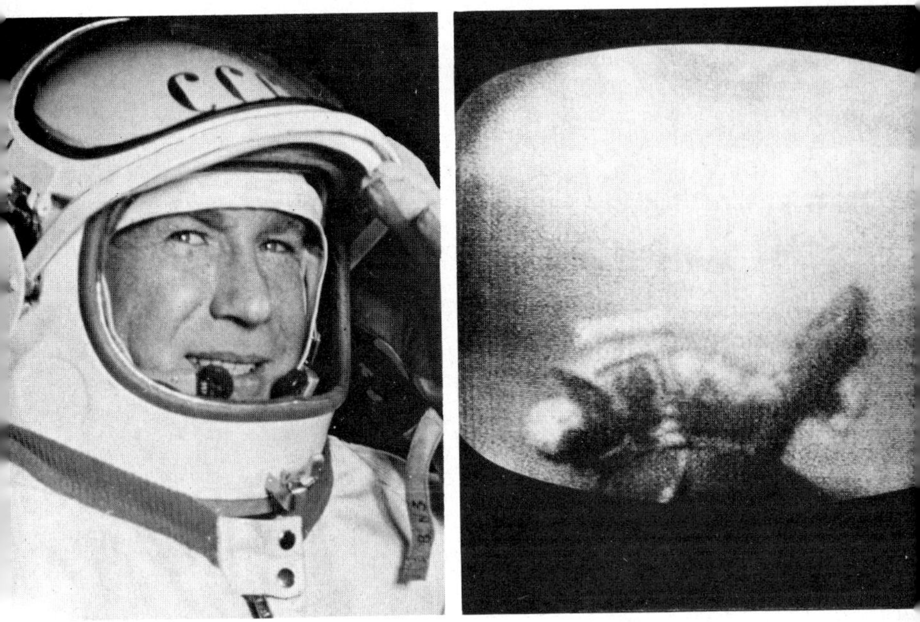

9. (*Above*) Two photographs of Alexei Leonov (born 1934) who, during the flight of Voskhod-2 became the first walker in space.

(*Below*) Leonov training in the air-lock of Voskhod-2. At the bottom of the photograph the hatch opening may be seen. Behind Leonov is the closed hatch giving access to the capsule.

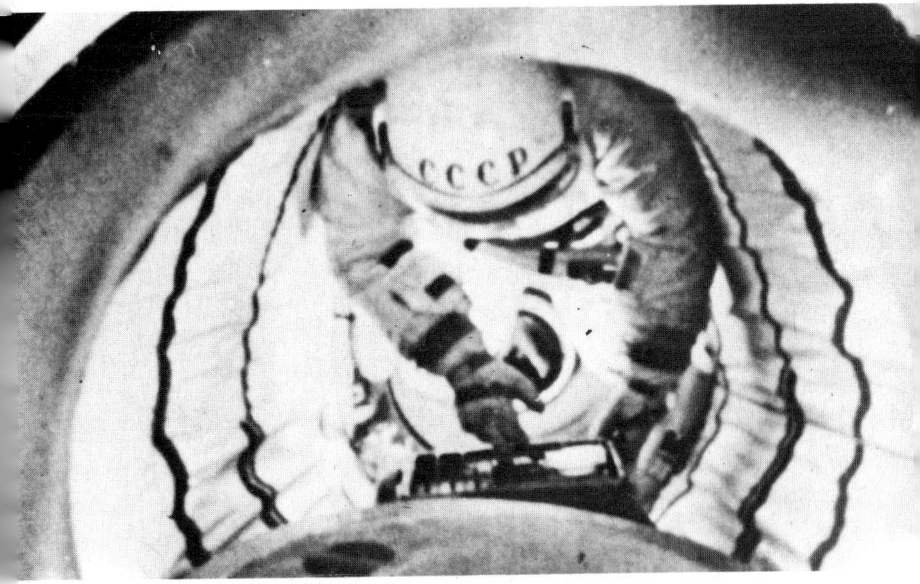

10. Leonov in his special space suit, floating like a human satellite around the earth. (*Photomontage*)

11. (*Above*) A representative of Russia's third generation of spaceships: Soyuz. *Left* the egg-shaped workshop, here shown with an experimental linking tunnel. *Centre* the command module, approximately cone-shaped. *Right* the engine section with two large solar cell panels.

(*Below*) Two Soviet cosmonauts who have since died. *Left* Vladimir Komarov with his wife and daughter. He lost his life in Soyuz-1. *Right* Yuri Gagarin, who crashed on a flight in a Mig-15.

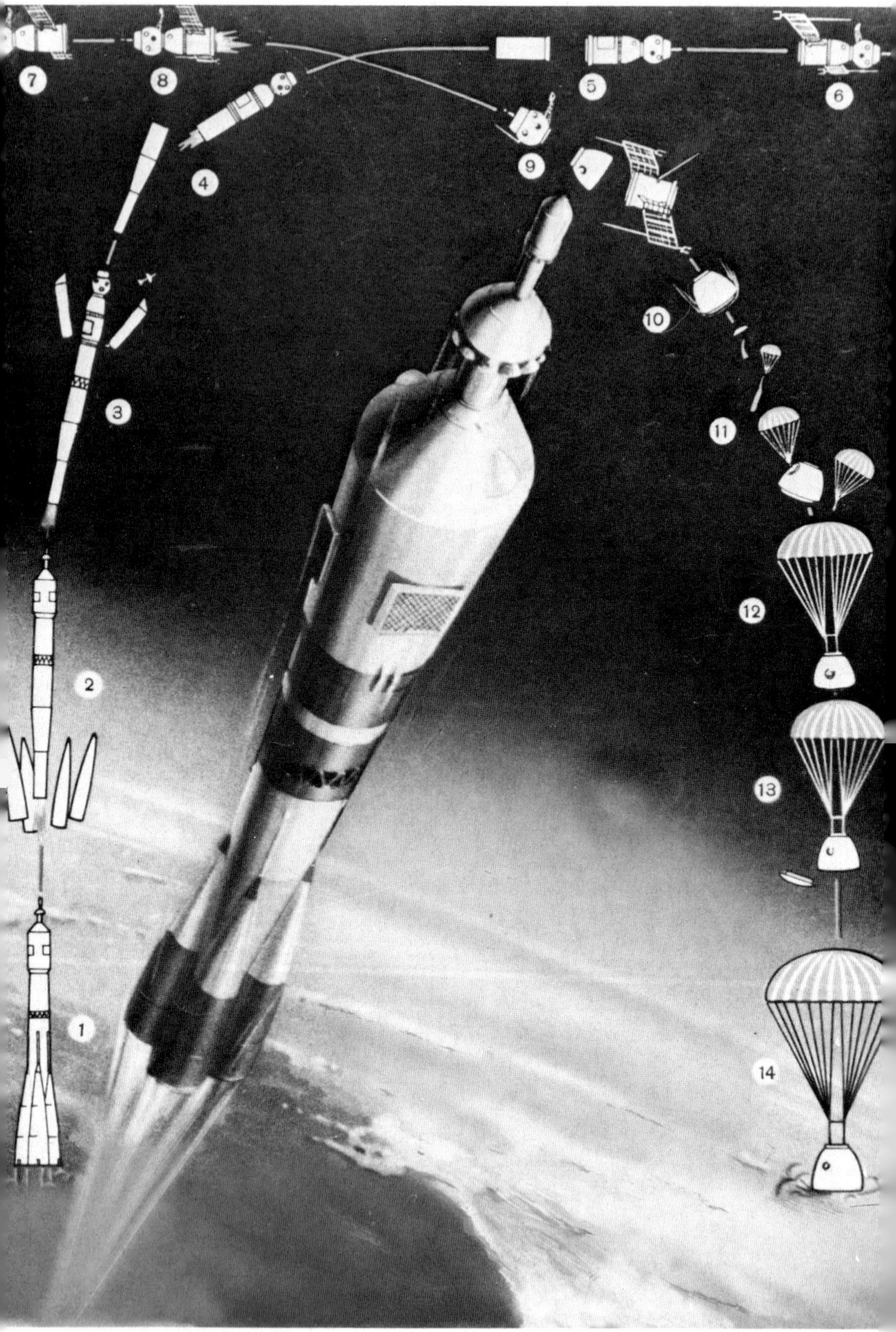

1
2
3
4
5
6
7
8
9
10
11
12
13
14

flight *round* the moon in 1968. The spaceship was to be launched by an improved version of the rocket which had previously put the 12–17-ton Proton satellites in orbit round the earth. Sheldon expressed the opinion that this flight would be preceded by one or more experimental flights of unmanned capsules.

It looked as if Sheldon was to be proved right. On March 2, 1968 Zond-4, which had space for a crew, was launched, but it did not get into moon orbit. It was followed on September 15 by Zond-5, the first spaceship to return to earth after a flight round the moon, although it was decelerated somewhat too rapidly. Zond-6, launched on November 10, appeared to be a dress-rehearsal for a manned flight round the moon. This spaceship was decelerated gradually and made a safe landing on Russian soil.

Exactly two days after the start of Zond-6 NASA announced that the next Apollo operation would not be an orbit round the earth, as originally stated, but a moon orbit. After this, Russian Zond flights suddenly became less frequent. Nearly a year later another Zond was launched and after a further year, another. But the spaceships remained unmanned. . . .

12. *Left* Launching and recovery procedures of a Soyuz. 1: Blast-off; 2: boosters are jettisoned; 3: protective nose-cone is discarded; 4: third stage is fired; 5: spaceship in orbit, separated from third stage; 6: solar panels and antennae deployed; 7: first stage of the re-entry procedure; 8: retro-rockets are ignited; 9: work compartment and engine section are separated from the command module; 10: deceleration in the atmosphere; 11: parachute hatch opens, drag parachute deployed; 12: drag parachute released, main parachute deployed; 13: heat shield jettisoned, releasing landing rockets; 14: landing rockets ignited at 1 metre above surface, so that the module makes a very soft landing. The rockets are fired by means of an ignition probe which is 1½ metres long.

SPADEWORK

Co-operation with the Germans – Kibaltchitch designs a manned rocket – Tsiolkovsky lays the theoretical foundations – The first laboratories – GDL and GIRD – Korolyov: rocket-planes and rockets – Stalin Organs

Misinterpretation

"We've got the wrong Germans!"

Thus exclaimed an American general a few hours after the launching of the first Sputnik, when the "good" news of the world's first earth satellite became known in the United States.

He could hardly have been more wrong. Although his spontaneous reaction was certainly typical of the psychological shock which the United States had just suffered, it showed a total lack of understanding of the development of space travel in the Soviet Union. Certainly, German rocket scientists—chiefly men who in Hitler's Reich had been working on the V-weapons—had gone to work in America as well as in Russia after the war. There was, however, one important difference: the Germans had stayed in America, whereas in the Soviet Union they worked for only a few years. With a few exceptions they were taken back across the frontier near Frankfurt on the Oder towards the end of 1953.

It is true that the Russians had made good use of the knowledge which the German rocket experts had gained with the development of the V-2 and other missile weapons. On the other hand, the Russians themselves had already made a fair amount of progress in the field of rockets prior to 1940. The notorious Stalin Organ, a mobile rocket battery which had caused panic among the German lines on more than one occasion, was only one indication among many of this progress. And as far as theoretical foundations for space travel were concerned, the Russians had by then achieved more than any other nation—including Germany.

In fact, in the twenties the Germans were making full use of the work done on space travel by Russian theorists. They were able to do so as a result of the fairly intensive co-operation between Russian and German specialists of which even now little if anything is known. These exchanges were a logical outcome of the extensive contacts which at that time existed in the field of aircraft construction. Around 1925 Junkers, for instance, built 200 planes a year in a factory not far from Moscow! In this connection *Pravda* and *Izvestia* wrote on December 16, 1926: "We give foreigners concessions for building factories and establishing companies whose products are for our own consumption; why should we forbid them—in fact why should we not urge them—to build industries for the defence of our country?" Neither of these papers neglected to mention that Junkers were indeed building planes for the Soviet Union. "The USSR has no objection to foreign engineers being used to strengthen its defences against foreign imperialists."

Valier

At that time the few space travel enthusiasts in Russia were in a better position than those in Germany or elsewhere in the West. Since Konstantin Eduardovich Tsiolkovsky (1857–1935) had laid the theoretical foundations for space travel, many people in his country as well as abroad had become enthusiastic about the idea that worlds beyond the earth might be reached. But whereas in the Soviet Union like-minded spirits soon made contact with each other and could thus work out some sort of programme, however modest, the subject elsewhere remained for a long time a matter of purely individual interest. The Soviet state, moreover, made some financial contribution—albeit a small one—to encourage research in this field. Lenin himself took an interest. In other countries rocket enthusiasts had to depend on rich industrialists, who were rarely prepared to invest money in "those lunatic rockets". If they did so, it was only as a publicity stunt.

One man in the West who experienced these difficulties was the South Tyrolese Max Valier. In the summer of 1924 he had made contact with Hermann Oberth, later to become Wernher von Braun's teacher and one of the first men in the

West to propagate the idea of space travel seriously. Like Tsiolkovsky, Oberth was above all a brilliant theorist, whereas Valier veered towards a practical approach. Valier soon formulated a practical programme which would provide the necessary experience with rockets in the shortest possible time. He wrote to Oberth: "Your idea is first to build and launch a small experimental rocket. This will cost about 50,000 marks. Even if the flight succeeds, the rocket will be lost . . . and you talk of sending rockets to the moon, at first unmanned and later manned! I believe we must go about it in a different way. In the first place we must completely master the technical part of the rocket, that is, the engine. We should not start firing immediately, not even small, experimental rockets. Let us, at our leisure, build such an engine into a testing frame, secure it carefully, and then fire it to see how the exhaust works. In this preliminary stage the machine need not have the shape of a rocket—it can have any arbitrary shape, depending on the position of the fuel tanks.

"Once we have mastered the purely technical aspects, without having launched anything in the air, and when the supplies of alcohol, water and liquid oxygen function well, we shall construct something that looks like a rocket. We shall mount the entire thing on a railway-truck and fire the rocket. Because of the recoil, the truck should start to move. As weight and friction will be known factors, we shall be able to form a practical judgment of the capacity of the rocket engine and we shall be able to calculate exactly how much of the energy developed by the fuel can in fact be used. . . ."

After these tests, according to Valier, the engine would have to be built into an aeroplane. If these experiments did not disappoint, the engine should be mounted in a real rocket and then, said Valier, "we shall gradually penetrate the universe."

The advantage of Valier's method was that no large sums of money were needed before a start could be made. Moreover, he thought that his initial experiments would create so much interest that industry would be able to get involved in subsequent experimental flights, which would require large-scale investment.

In September 1925 Valier met the management of the Junkers aircraft factory. The result was hardly sensational. The Junkers people showed interest in the project, but said they could take no part until a suitable rocket engine had been constructed and tested.

Valier later disclosed that a number of people had been present at the conversation whose names he did not know at that time. They were, in fact, Soviet rocket technicians. The Russians emphatically declared that *they* were prepared to give Valier financial support in the development of a reliable rocket engine. This amounted, in fact, to a Russian offer to Valier to join their ranks, but he did not accept it. He still believed he would be able to create sufficient interest in his own country. He was to be disillusioned in many respects.

However, the Russians had listened carefully to Valier's exposition and they turned it to good account. Men like Sergei Pavlovich Korolyov, later the mysterious "chief constructor", in fact began working along the lines proposed by Valier; as a result the Russians developed the rocket-plane before the rocket. In America the reverse happened.

In that period the Soviets already had adequate facilities for developing rockets, practically as well as financially. They offered Valier "excellently equipped laboratories". He did not respond. He even refused an invitation to visit the International Space Travel Exhibition organised in Moscow in 1927, although he did send the Russians, at their request, his plans for the development of rockets and spaceships. Needless to say they accepted them eagerly.

In the course of the next few years Valier managed to carry out experiments with rocket cars, financially assisted by Fritz von Opel. Occasionally a solid fuel rocket engine was successfully mounted in a glider. These experiments, too, were attended by observant Russian experts. The Germans were, in fact, invited to experiment in the Crimea with rocket-powered gliders.

Tsiolkovsky

Meanwhile, however, the Russians had already gone quite some way in carrying out a considerable programme, of which little or nothing was at that time known in the West. As regards "co-operation" with the Germans: this in practice proved to be a rather one-sided agreement, from which the Russians benefited most. Although they soon made much greater progress with their experiments than the Germans, they were naturally always interested in any new German development. Russian rocket technicians had every opportunity to keep themselves informed of German advances in this field almost up to the beginning of the Third Reich. This obviously raises the question of why the Germans showed their hand in this fashion, but the so-called "exchange of information" in the field of rocket development was merely a side issue of the Treaty of Rapallo, a military agreement between Germany and the Soviet Union. Shortly after Hitler assumed power the exchange was stopped.

By then the Russians had more than half a century of research and development behind them and had thus laid the foundations for their subsequent successes in the field of space travel, which would give them the lead for some time to come.

This development had, in fact, started as early as 1850—more than a century before the start of the first Sputnik. At that time General of Artillery Konstantin Ivanovich Konstantinov, was already engaged in an intensive study of solid fuel rockets. However, he did not confine his attention to the various kinds of fuel but also studied the external shape of the rocket, its control and launching apparatus. In 1881 Nikolai Kibaltchitch published a plan for a manned solid fuel rocket. At that time, however, his plans created little interest, for he drew his designs in the Citadel of Peter and Paul in Petrograd (later Leningrad), where he was imprisoned for making the bomb which killed Tsar Nicholas II. A few days before he died, Kibaltchitch requested that his ideas should be investigated and tested, but it was not until the time of the 1917 revolution that his notes, which were filed in the archives, were published. It was a long time before Kibaltchitch was publicly honoured for his scientific work, namely at the celebrations, held in Moscow's Red Square, in honour of Gagarin's flight, when Prime Minister Khrushchev said:

"Today, as we are standing next to the first man to penetrate into the cosmos, we must not fail to remember the Russian scientist and revolutionary Kibaltchitch, who dreamt of flights in space and was condemned by the Tsarist government. . . ."

Fortunately there were other men in Tsarist Russia who

thought like Kibaltchitch. Each of them contributed something to the development of rockets, but none of them made such an immense and valuable contribution as Konstantin Eduardovich Tsiolkovsky, the deaf schoolmaster from Kaluga. Almost single handed he initiated a century of research into space travel in the broadest sense of the term. Tsiolkovsky worked out the fundamentals of space travel and laid down the lines along which its development should take place. He was largely unrecognised in his day which is hardly surprising for many of his projects appear sensational even now.

It is amazing that a man like Tsiolkovsky, who led a solitary existence for most of his life and maintained few contacts even with like-minded people, could become the "father of space travel".

He was born in 1857 at Ijevsk, the son of a woodcutter. As the result of a serious illness he became almost totally deaf when he was ten years old. His mother taught him as much as she could. It was soon apparent that Konstantin Eduardovich was a profound thinker. He read voraciously, mainly scientific works. Konstantin was only twelve years old when his mother died. Her death affected him deeply and as a result he isolated himself even more from his comrades. One day he accidentally got hold of a balloon and this became his first contact with the rocket principle. Why does a balloon move if I untie the opening? the boy asked himself. The question did not leave him alone until he had found the answer.

Tsiolkovsky realised that the air in a closed balloon exerts the same pressure on all sides. The pressures in different directions thus cancel each other out and the balloon does not move. However, where there is an opening, the pressure is removed, so that the pressure on the opposite side, in this case the front of the balloon, is no longer counteracted, and consequently the balloon moves forward. At the same time it became evident that the balloon would fly equally well in a vacuum as in the atmosphere. In fact, even better, for in a vacuum there is no resistance. Contrary to what is often thought, the balloon therefore does not push itself off from the surrounding air.

This gave Tsiolkovsky the idea of a liquid fuel rocket engine. As long as the open balloon contained gas under high pressure, it would continue to fly. This meant that there would have to be a constant supply of new gas to replace the exhaust gas. Naturally the pressurised container would have to be a metal one instead of a rubber balloon and the gas would have to be conveyed in liquid form, otherwise the volume would be impractically large.

Tsiolkovsky's father was aware of his son's ability and he sent the studious boy to Moscow, where Konstantin spent most of his time in the great libraries. He also wanted to do practical work, but for this he needed apparatus and chemicals. To be able to buy these he sometimes half-starved himself for days on end. The money thus saved was invested in apparatus and materials. As a result his health, never very good, could hardly be expected to improve. Eventually Konstantin's father made him come home. In 1882 Tsiolkovsky became natural history master at a school in Kaluga. His spare time was increasingly devoted to the study of space travel, although his activities met with little appreciation among his fellow villagers, who regarded him as an eccentric. And not without reason. Not only did Tsiolkovsky, in his numerous articles, publish ideas which at that time seemed little more than idle fantasies, he also lived a very solitary life. Possibly his attitude was partly the result of his bad hearing which made it difficult for him to establish contact with other people.

Tsiolkovsky also worked out a mathematical elaboration of the rocket principle, and the famous "basic formula of Tsiolkovsky" is applied to this day. He proved that a rocket is the only method of penetrating space; it can carry its own oxidant and fuel for combustion, as opposed to aeroplanes, which either thrust off from air (propellers), or at least require air for their engines (jets). Tsiolkovsky propounded all this at a time when even aeroplanes were hardly thought of. He moreover constructed, among other things, a model for the first dirigible airship and built the first wind-tunnel for aerodynamic experiments with his models. Making real rockets was beyond him, for he received no encouragement whatsoever from the Tsarist government. Nevertheless, he has never been surpassed as a theorist and may rightly be regarded as the father of space travel. He propounded that the multi-stage rocket was the only method by which the necessary cosmic speeds could be achieved with existing sources of energy. Such a multi-stage rocket could consist of a number of stepped-up rockets—the method now generally used. It would also be possible to use a "cluster" of booster rockets to be jettisoned one by one. Tsiolkovsky provided a concrete example. Place four interlinked rockets in a row. Fire them off together. When half the fuel carried is used up, the fuel in the two outer rockets is transferred to the two inner ones. The outer rockets drop away. When the remaining fuel is reduced to half, the residue flows from one rocket to the other, and the empty rocket is jettisoned. The remaining rocket finally reaches the desired speed. This extremely original, but complicated theory has never been put into practice. Nevertheless, Russian engineers later used the principle of parallel rockets in the design of the standard launching rocket (used for launching manned and unmanned spacecraft), consisting of a central stage and four boosters which are jettisoned when empty.

Initially Tsiolkovsky's activities were not widely known, but there was very much criticism of his published work. There were sceptics who thought that the schoolmaster from Kaluga had given too free rein to his imagination. Tsiolkovsky had a valid answer to this: "There was a time—and not so long ago at that—when even famous researchers and theorists ridiculed the idea that one day we would know the composition of heavenly bodies. That time is past. It seems to me that the conception of direct reconnaissance of the universe might at the present time seem even more lunatic. To set foot on an asteroid, to pick up a rock from the moon, to construct mobile stations in the ether, to create inhabitable belts around the earth, the moon and the sun, to study Mars from a distance of a few *versts*, to land on its satellites, or even on the planet itself—all this might sound absurd. Nevertheless it will be by means of reaction apparatus that a great new era of astronomy will be opened up, an epoch in which the skies will be examined in depth. Do not let us be deterred by the immense powers of gravity. It does not surprise us that a cannon ball flies at 2 km a second, so why are we frightened at the thought that it might reach a speed of 16 km a second, to leave the solar system forever and, having defeated the powers of attraction of the earth, the sun and the entire solar system, to penetrate into endless space? Is the gap between 2 and 16 really so great? After all, sixteen is only eight times two!"

At the time of the revolution Tsiolkovsky was 60 years old. He had never received the slightest support from the Tsars, but when he hopefully approached the Soviet government, it was a different story. In 1919 he was made a member of the Academy of Sciences and between 1925 and 1932 no less than sixty of his writings were published. The state granted him a modest pension.

Meanwhile Tsiolkovsky kept an eye on developments abroad and thus learnt about Max Valier's plans. He corresponded with Hermann Oberth, later Wernher von Braun's teacher, and with Ari Sternfeld, who later became the chief theorist in the Russian space programme. To the latter Tsiolkovsky wrote in 1935, the year of his death, that at one time he had thought that it would be centuries before his plans could be realised. Recent developments had convinced him that a few decades would suffice. "The efforts and sacrifices which our Soviet government is making for the development of the industry and for diverse research will, I hope, justify my expectations."

The first laboratories

At the turn of the century there were, apart from Tsiolkovsky, several other Russians whose remarkable achievements in the fields of physics and chemistry benefited the young science of space travel. As so often happens in history, there were like-minded people who, independently of each other, conceived ideas which were to have a considerable influence on subsequent decades. Obviously the time was ripe. There was, for instance, D. I. Mendeleyev, who in 1869 published his famous *Periodic System of Elements*, which for the first time clearly showed the mutual cohesion of chemical elements. At the same time this provided insight into the manner in which some combinations of elements react, while others do not.

This knowledge was of inestimable value for the development of the rocket, which is propelled by chemicals. Also of great importance was the work of I. V. Merschersky, who in 1897 introduced his "Dynamics of a point with variable masses". This was exactly what rocket builders needed. As fuel is used up, the weight of a rocket decreases in flight, but as the thrust remains the same, its acceleration increases. At the beginning of the present century the theory of aerodynamics was taught in many institutes and universities, particularly in Moscow. Consequently it need not surprise us that the Russians gained the lead in aircraft construction as well. This kind of knowledge was of equal importance to rocket builders, for during a considerable part of its propelled flight the rocket moves through the earth's atmosphere.

Because of the First World War and the chaotic conditions created by the February and October Revolutions which followed it, very little happened in Russia in the field of space travel between 1914 and 1922. Not until the twenties was any further headway made. It was during that period that "co-operation" with the Germans was established as a result of the Treaty of Rapallo in 1922.

Of course it was much simpler to make solid fuel rockets than missiles using liquid fuel, so a beginning was made with the former. In 1921 two engineers, Nikolai Tikhomirov and Vladimir Artmeyev, founded the first research laboratory for these rockets. In 1927 this modest establishment was transferred to Leningrad under the name Laboratory of Gas-dynamics of the Revolutionary Military Council (GDL). The laboratory came under the supervision of M. N. Tuchatsyevski, who was also in charge of the Council's armaments bureau. Tichomirov continued to be in charge of the day-to-day management. Three years later the *Gruppa Myezplanyetnich So-obtsyenii* (Group for Inter-planetary Communications) was established in the famous Zyukovski Academy of Aviation in Moscow (founded in 1919). In May of the same year the name of this organisation was changed to *Obsyetstvo po Izutsyeniye Myezplanyetnich So-obtsyenii* (Society for the Study of Inter-planetary Communications). It was chiefly on the initiative of this very active society that in 1931 the GIRD was created, the *Grupa po Izutsyeniye Reaktivnovo Dvizyenia* (Group for the Study of Reactive Propulsion). The establishment of this group, which had branches in the capital as well as in Leningrad, in effect initiated the practical development of both rockets and rocket- and jet-planes in the Soviet Union.

It was in GIRD that the modest, but nonetheless important precursors of the Vostok and Soyuz launching rockets were born. It was also in GIRD that the first ideas of a young rocket enthusiast, Sergei Pavlovich Korolyov were realised. Korolyov was to become the "chief engineer", the man whose name in the era of practical spaceflight would be one of the most closely-guarded secrets in Russian science, and that until his death in 1967.

In Moscow GIRD was headed by A. F. Tsander (1887–1933), who as early as 1924 had produced a highly revolutionary plan for a rocket-plane. The remarkable thing about this machine was that, when it left the atmosphere, it would use its own wings, which were by then superfluous, as fuel. The project was never carried out and in the years that followed Tsander occupied himself mainly with working out the theoretical foundations of rocket technology.

The first real rocket engine was built and tested by the GDL in Leningrad in 1931. It was the ORM-1 (*Oputnii Reaktivnii Motor*—experimental reaction engine), burning petrol and liquid oxygen and producing a thrust of 20 kg, which for that time was quite considerable. At the same time Tsander in Moscow built the OR-1, which produced a 5 kg thrust.

In April 1932, exactly 29 years before Gagarin's flight, Korolyov was put in charge of the first experimental Soviet rocket design centre in Moscow, which embraced various existing organisations. Obviously the Soviet government by that time recognised the importance of rocket development.

Tsander continued developing rocket engines, and another group, headed by M. K. Tikhonravov, developed the first Soviet rocket, the GIRD R-09, which was launched on August 17, 1933. It was as yet a small rocket, 2·5 metres long, with a weight of 19 kg and a thrust of 50 kg. On November 25 of the same year, the GIRD R-10 was launched.

In December 1933 a further significant reorganisation took place: all research was co-ordinated in the *Reaktivnii Nauchno-Issledovatyelkii Institut* (Rocket-science Research Institute), or RNII. Kleimenov became its director and Sergei Korolyov its deputy-director. Up to the Second World War the RNII was to remain the chief institute for rocket development and production. The famous Katyushas ("Stalin-organs") originated here as well. Rocket engines with a thrust of up to 320 kg were built. On April 24, 1936 the Aviavnito (length 3 metres, thrust 300 kg), at that time the largest Russian rocket, was fired. On August 15 a similar missile reached a height of 2400 m.

Rocket-planes as well as rockets were developed by RNII. This department was under the control of Korolyov, himself originally a pilot. His first effort was Project 212, a small, unmanned rocket-plane, like the German V-1, with a range of 50 km. The machine, which had a wing-spread of 3 m and weighed 210 kg, flew in 1939. A manned rocket-plane—also designed by Korolyov—flew on February 28, 1940. This was the RP-318, piloted by V. Fyodorov. The machine was towed to 2600 m by a conventional plane. At this height it was released and the rocket engine, designed by L. S. Dushkin, went into action. Within six seconds its speed rose from 80 to 140 km per hour. Fyodorov reached a height of 3 km and landed safely.

Meanwhile the Russians were by no means blind to the developments which had taken place in this field abroad. Their contacts with the Germans had already yielded a considerable amount of information, eagerly collected by the recently founded BNT-MAP (*Buro Novoi Techniki Ministerstva Aviatsionoi Promishlennosti*—Bureau for Modern Techniques of the Ministry of Aviation-Industry). Naturally their activities were intensified during the war, and the Russians knew better than anyone else what was being done in rocket engineering abroad. At the end of the Second World War, therefore, they knew exactly where they stood: the theoretical foundations had been laid and excellent results had been achieved with small rockets. What was lacking was a system for producing large, powerful rockets (such as the German V-2), which were of real importance for military purposes as well as for space travel. This situation was to change rapidly in the next few years.

THE SUPER-ROCKET MYTH

Germans and V-2's to Moscow – Gröttrup and his team – The first Russian long-range rockets – The Sputnik conception – The simple solution: clustering – Super-rocket

Germans to Moscow

It was the 22nd of October 1946. In East Germany the Soviet military had, on orders from Moscow, begun—sometimes in a rather high-handed manner— to "recruit" German scientists and technicians who had worked under Wernher von Braun on Germany's Vergeltungswaffe-2. Admittedly, they no longer had first choice. Immediately after the end of the war quite a number of rocket- and nuclear experts, including von Braun himself with a small army of his best collaborators, had made post-haste to America. The Russians meanwhile had laid their hands on about a hundred fully or partially assembled V-2's, plus the necessary blue-prints of more promising projects, and Moscow considered that it was in any case advisable to bring the remaining German rocket technicians to the Soviet Union. Their Russian "colleagues" would thus be able to acquaint themselves much more quickly with the level attained by German rocketry at the end of the war.

The Kremlin had not taken this decision without opposition. Was it really necessary to bring the Germans to Russia? Would it be possible to learn anything of what the Germans had achieved in the last few years? One of the opponents of the idea was Sergei Pavlovich Korolyov, later "the chief engineer", who remarked: "We ought to have more faith in ourselves." Others, such as Marshal Beria and Colonel Serov, inspired by the declarations of Joseph Stalin, thought that it could in any case do no harm to admit the Germans. Of course one need not let them take an active part in the construction of new rockets. Many Russian party-leaders shuddered at the idea of allowing future rocket development in the socialist fatherland to depend very largely on a number of former enemies. Finally it was agreed to bring the Germans to Russia, but to make them work in strict isolation. Everything they did would be closely watched by the Russians, until nothing further could be learnt. Then the Germans would be sent home.

Certainly the Russians could do with some German help, for thanks to an American blunder they had captured a large number of V-2's and machinery for their manufacture. The rockets had been made in an underground factory at Nordhausen. Immediately after the end of the war the Americans occupied this factory for ten weeks, but on the strength of some vague instructions they abandoned the whole of the valuable establishment to the Russians, who, within hours of the Americans' departure, promptly started to pack the contents. Under the command of Georgi Malenkov the booty was sent to Moscow. Not only complete V-2 rockets, tools and machinery, but also the designs for the A-9/A-10 rocket had fallen into the hands of the Russians. The A-9/A-10 was a two-stage rocket designed to wipe out New York. Fortunately the war ended before the Germans were able to fire it. Only the second stage, the A-9 (an improved version of the V-2) was ready by 1945. The A-10, a giant with a 180-ton thrust, only existed on paper.

Although it is certain that the Russians have never built an A-9/A-10 as such, there is no doubt that they borrowed many of the project's elements in the construction of their first medium-range and intercontinental rockets. Their equivalent to Wernher von Braun was Helmut Gröttrup, a man who is now peacefully working in Germany on computers. In Peenemünde Gröttrup had been engaged in the design of electronic equipment for the V-2, and immediately after the war he was in West Germany. By then the Americans had already taken von Braun to the United States, and with him a considerable number of his fellow-countrymen who had worked in the field of rocket construction. There were, however, many technicians for whom the chances of getting going in their chosen field in the New World did not seem very good.

Moreover, in the prevailing uncertain conditions, many people elected to remain in Germany, either in the west or in the east. Gröttrup finally decided to go to East Germany, since he knew that the Russians had captured the lion's share of the material. He thought that he would be able to continue his work on rockets in the Soviet-occupied zone of his country, and many German technicians thought likewise. Together with 200 others, Gröttrup sought refuge in the east. They were indeed given the opportunity to start on a limited production of the V-2 in the *Zentralwerke*, under the command of a Russian, General Gaydukov. Within a year this factory had 5500 workers. The Russians guaranteed that no one would be taken to Russia against his will. Germans could continue to work on German soil. But eventually the Kremlin changed its mind. Stalin instructed General Ivan Serov to move all the German technicians to the Soviet Union, and that is what happened.

The Russians did their work thoroughly but with scant subtlety. During the period from the 12th to the 16th October 1946, 6000 German technicians of all grades, together with their households, were "loaded" into trains. Relations, wives, children, grandmothers, servants—they all went: 20,000 people in all.

A few days later they arrived in Moscow. Gröttrup took the very first available opportunity to protest officially, in the name of the *Zentralwerke* for whom he had worked in East Germany, against what he considered their inhuman treatment. A few weeks later his protest was returned together with a comment against which there was little argument. The Russian who translated the reply made it clear that in the first place the *Zentralwerke* no longer existed, so that it was not possible to protest in the name of this organisation. Moreover, the spokesman continued, the Potsdam Treaty enabled the Soviets without further ado to recruit Germans for restoration work. Gröttrup should realise that, if he continued to protest, he and his men would be transferred to the Ministry of Mines. In the face of this threat there was little Gröttrup could say.

The Russian attitude was right insofar as the Soviets had indeed lost a number of their rocket bases as a result of the war and could thus reasonably order the Germans to repair the damage.

Gröttrup's men formed a very heterogeneous company. Only a fraction of his colleagues, 200 in all, had actually worked in Peenemünde. Most of the other technicians came from other German industries. There were men from the rocket-testing station at Lehesten, radar and electronics experts from Lorenz, Siemens and Telefunken, and a considerable number of specialists from Askania, a company engaged in the manufacture not only of compasses, but also of theodolites and measuring apparatus suitable for measuring rocket trajectories.

In any case there was more than one German group. Dr. Wilhelm Fischer, an expert in the field of rocket guidance, who had been taken prisoner by the Russians as early as June 1945, was put in charge of a group which was to spend the rest of its sojourn in the Soviet Union behind barbed wire.

"Russian" V-2

Initially Gröttrup and his men stayed in the neighbourhood of Moscow, where for eighteen months they were allowed to continue their work on the V-2. This group enjoyed slightly more freedom but little more comfort than Dr. Fischer's. The Gröttrups lived in a "dacha", a country house, and had the use of a chauffeur-driven car. In her book *Die Besessenen und die Mächtigen im Schatten der Roten Rakete* (Slaves and Masters in the shadow of the Red Rocket), Mrs. Irmgard Gröttrup writes that she was occasionally able to go shopping in Moscow. The Gröttrups and the rest of the German group were temporarily moved to the first Russian launching site for long-range rockets, a camp consisting of army tents, caravans, a number of experimental gantries and a launching pad. This base was situated in the steppes, 200 km east of Stalingrad.

It was here that, on October 30, 1947, for the first time in the Soviet Union a German-built V-2 was launched. Subsequently "Russian V-2's" also started from here—rockets with components made in Germany, but assembled in the Soviet Union. Some Russian technicians even proclaimed in all seriousness that the "Russian" rockets were better than the German!

Gröttrup and his colleagues did not have the opportunity to carry out actual launchings for long. The Russians soon took over and the German group was transferred to an island in the region of the sources of the Volga, 350 km northwest of Moscow. The island, at one time called Gorodomlia, is situated near the village of Ostyakov in the district of Kalinin. Here, the Germans in collaboration with a large number of Russian technicians, tested improved rocket engines designed under the leadership of Dr. Jochen Umpfenbach. The great shortage of component parts at times inspired the Germans to brilliantly simple constructions. One of these was the Pobyeda, a greatly improved version of the V-2, which had little in common with Hitler's original weapon. The Pobyeda had a straight, cylindrical fuselage and a removable extremity for the payload. In spite of its greatly simplified form, the Pobyeda (a later version became the T-1) had a range of 910 km. The Germans called this project R-10. They themselves never saw this rocket fly. As soon as the plans had reached a certain stage, they were sent to Moscow. Later Gröttrup and his team were often requested to develop other constructions.

Strategic rocket

The last great project initiated by Gröttrup was the R-14, a strategic rocket with a range of 3000 km and a payload of 3 tons. The R-14 was conical in shape and had no fins; it was powered by a Gimbal-mounted, 100-ton engine, which could be used to change course.

Early in 1951 the island was, as it were, invaded by Russians. The Germans gradually became a minority, and more and more of the work was taken out of their hands. The Germans believe that the Soviets used the R-14 design in the development of their first intercontinental missile, but they were never given the slightest indication that the R-14 plans were being carried out.

During the last years of their stay in the Soviet Union the Germans were made to work on a number of small projects, including a computer. Because of uncertainty concerning their possible return to Germany, they lived under constant strain. The Russians probably intended that the *Nyemyets* should forget as much as possible of what they had been doing in the Soviet Union. Gröttrup's team returned to Germany at the end of 1953. Other groups were made to stay longer and spent many years doing nothing. The last group did not return till 1958. Only a few Germans preferred to stay in Russia.

Although the German technicians who were sent to the Soviet Union did not play a decisive role in the development of Russia's rocket and space programmes, it must be said that they undoubtedly gave the Russians a start. True, the

Russians themselves had the necessary experience in the field of rocket development, and theoretically they were in fact very well informed; but up to the end of the second world war they had not got round to the production of large rockets.

Obviously the Russians rarely mention the part the Germans have played in their country. An exception was Khrushchev who, in the course of a speech in Minsk on January 22, 1958, declared somewhat excitedly:

"If the Germans have helped the Russians, why aren't they helping the United States? After all, it was the American army who got hold of the chief constructor of the V-2 and took him to America, where he is now building rockets."

In a calmer voice he continued: "Indeed, a small group of Germans worked here. At the end of their contract they returned to Germany, and that is where they are working now."

The V-2

Although by the end of the war the V-2 was by no means perfect, it incorporated so many original constructive ideas, that the Germans may rightly be said to have made the rocket usable in spaceflight. It was hardly surprising that Russian experts, such as Sergei Pavlovich Korolyov, took a great interest in Hitler's former weapon. The V-2 was a truly formidable affair. The cigar-shaped body, fourteen metres long, had four enormous stabiliser fins, with rudders working in conjunction with steerable graphite vanes in the exhaust stream, so that the rocket's course could be corrected. This, of course, only applied while the V-2, which could reach a maximum height of 90 km and had a range of 330 km, remained in the denser strata. In the vacuum of space these fins would naturally have no effect. Their function was then taken over entirely by the exhaust vanes.

The propellant used for the V-2 engine was ethyl-alcohol, with liquid oxygen for combustion. The engine developed a thrust of approximately 27 tons, and the rocket weighed 13 tons. The V-2 could carry 750 kg of explosive. Since the weapon was inaccurate—its course could not be corrected and the fuel supply was cut off by a simple time mechanism—it rarely hit its predetermined target or even the surrounding area. It therefore caused relatively little damage, although no fewer than 4300 reached England.*

From a military point of view one of the chief advantages of the V-2 was the fact that it could be fired from a mobile launchpad. The Russians later adopted this idea.

Russian rockets

It would be inaccurate to say that the Russians worked solely on German designs when after the war they set up their own rocket programme, which was mainly of a military nature. On the basis of their experience with the GIRD rockets they developed more powerful versions, intended among other things for the study of the upper atmosphere. Their construction was and is carried out by the Moscow OKB (*Opitnoy Konstruktorskii Buro*—Experimental Design Bureau). In the field of rocket construction the Russians had plenty of technicians of their own, who were every bit as good as the Germans. But of course they were happy to make use of the advances which the Germans—because of Hitler's lust for power—had achieved within a short time.

The Russian V-2A and V-5V, designed for purely scientific research (such as experiments with animals who returned to earth in their capsules) were directly derived from the V-2. The first tests with these rockets were carried out in 1951. Their research into the effect of space travel on organisms were of great benefit to science. Animals were launched in vertical experimental flights to heights of up to 450 km. This research was applied chiefly to dogs, but rabbits, rats, hamsters and mice were also used.

Independently of the improvement of the V-2, the Russians were also working on other programmes. To begin with, they built their Pobyeda (Victory), which was capable of considerably greater performance than the German weapon. Nevertheless the Kremlin, in particular, was none too satisfied with this rocket. "This isn't what we want," Georgi Malenkov said in 1947 to Gregory Tokayev, a rocket engineer who later sought refuge in England. "We have improved the V-2, we

*The first V-2 fell on London on September 8, 1944. It was launched from Wassenaar, Holland.

have got further than Peenemünde in 1945, but it remains a primitive, blind and short-range weapon. Whom do you think it will frighten? Poland? Turkey? We don't want to frighten Poland! Our potential enemy is thousands of kilometres away. We must get on with the development of long-range rockets."

"*Abstract Sputniks*"

In the course of the above conversation Malenkov referred to Dr. Eugen Sänger's project for a rocket-plane which could fly halfway round the earth. Sänger "unfortunately" had ended up in France, not in Russia, but the Russians had got hold of his designs. The rocket-bomber was to be launched from a ramp nearly 2 km long. Although the Russians have never used the rocket-bomber as such, they adopted certain elements from the blueprints in the construction of their own long-range and inter-continental rockets, which, as is usual, were launched vertically and changed to a more or less horizontal course only after they had left the atmosphere. Perhaps it was the knowledge that the Russians had got hold of Sänger's design which led the French journal *l'Auto-Journal* to state in May 1961, that is, a month after Gagarin's flight, that the Russians used a launching rail 30 km(!) in length for their manned spacecraft. According to this magazine, the Russians had built a long railway in Mongolia on a 45 degree incline. Booster rockets were to be used to achieve maximum acceleration, and at the end of the railway the rocket engines were to be fired. By using this technique the Russians were said to be able to use rockets with hardly greater thrust than those of the United States, but which could carry much greater payloads into space.

Later, films of the launchings of Vostoks showed clearly that the French story was a complete fantasy. Obviously the writers did not realise that it is, on the contrary, essential to avoid the enormous air resistance which this technique would involve. The rocket must leave the atmosphere as quickly as possible. In any case the writers admitted that they only had one single indication on which to base their daring theory, namely, the fact that each take-off took place at the same angle of 65 degrees with the equator. On later occasions the Russians were to vary the angle, depending on the objective of the spacecraft involved.

As early as the end of the 'forties, some Russian rocket experts talked openly about realising Tsiolkovsky's dream of putting a satellite (sputnik) into orbit round the earth. One of these experts was Sergei Pavlovich Korolyov. But the Soviet rulers of that time, Stalin of course among them, considered this scheme of little use. Gregory Tokayev remembers that Lt. General Kutsevalov, at that time commander of the air force in Berlin, once remarked that American planes violated Russian air space and said: "This proves that the danger of a new war remains with us. It is therefore not the time for abstract sputniks." One adherent of the sputnik scheme—although only in an unofficial capacity—was Malenkov, as appears from a meeting in the Kremlin on March 14, 1947. Malenkov asked Tokayev whether any progress had been made in "the matter". "Yes, but not much," replied Tokayev. "Carry on," whispered Malenkov, "and let me know as soon as you have something worth while." Deputy Prime Minister Nikolai Voznesensky also agreed in principle that, simultaneously with the development of military rockets, work should be carried out on a plan for launching an earth satellite. In later years work on this project was to be controlled chiefly by the BNT-MAP (*Buro Novoi Techniki Ministerstva Aviatsionoi Promishlennusti*) and the OKB mentioned above.

The Simple solution: clustering

But first the rocket had to be made! Although a special committee was formed to study, and if possible execute, Sänger's project (one of the members of this committee was Mstislav Keldysh, now president of the Academy of Sciences), it never got off the ground. Nevertheless rapid progress was made with medium-range rockets, thanks largely to the support of the leaders in the Kremlin, who, unlike their American rivals at that time, regarded the rockets as one of the most important elements in any future war.

By around 1947 the Soviets had increased the V-2's range to over 500 km, with an accuracy of aim of 500 m, i.e. within only 0·1 per cent. When by 1950 the range had been further increased to 900 km, the rocket had undergone so many changes, that one might rightly speak of a new type altogether.

Outwardly the differences were particularly striking: it was no longer cigar-shaped, but cylindrical, which permitted more efficient production. The fuel tanks were larger, the fins much smaller, and the method of propulsion, too, had been amended. The Soviets called this rocket the T-1. This weapon is still in use in many units of the rocket forces.

The initial weight of the T-1 is 17 tons, four tons more than that of the German rocket. It develops a thrust of 35 tons (V-2: 27 tons). The T-1 uses kerosene instead of methyl-alcohol, but liquid oxygen is still used for combustion.

The Soviets succeeded in making the 25-ton engine, derived from the V-2, a particularly reliable apparatus. It was therefore not surprising that they used this engine for their first inter-continental rocket, which was announced by Khrushchev in 1957. The same rocket was to be used that year for launching the first sputnik and in the years that followed this standard work-horse was to send the most diverse kinds of artificial satellites and manned spacecraft into the cosmos. The central propulsive element of this rocket is the RD-107, a combination of four 25-ton engines. Fuel is supplied by one central turbine. Two small swivel-mounted engines have been added to make the whole thing steerable. The graphite rudders used in the V-2 were thus dispensed with, although they are still being used by the Soviets in smaller rockets. The RD-107 engine (the Russians call it an "engine", although in fact it is a combination of four large and two small engines) was developed by the OKB in the period 1954–57. Its construction is simple. It has everything which is absolutely essential, but nothing else. It runs on a combination of liquid oxygen and kerosene. These components are combined in a so-called pre-mixer compart-ment, from where they are injected through a large number of small jets into the combustion chamber, where the mixture is electrically ignited. The launching rocket of the Vostoks (RNV: *Rakyeta Nosityel Vostok*) has four RD-107 units. A slightly different version (with four steering engines), the RD-108, forms the central second stage. The first stage of the rocket consists of four separate rockets assembled round the central stage (see drawing pp. 66–67). A single 25-ton engine is built into the third stage.

Fig. 5. Soviet launching rockets.

1. V-2A, used for launching experimental animals. Length 20 metres.
2. RNK (Kosmos launching rocket), used for sending up lightweight scientific satellites in the Kosmos series. Length 33 metres.
3. Launching rocket for the first Sputniks. Length 32 metres.
4. RNV (Vostok launching rocket). Based on the Sputnik launching rocket. Used for launching manned Vostoks and—in several variations—for launching Voskhods, Lunas, Veneras and Molniya communication satellites. Length 38 metres.
5. RNS (Soyuz launching rocket). Used for launching Soyuz spacecraft. It has an improved third stage, but is otherwise in principle similar to the RNV. Length 49 metres.
6. RNP—Proton launcher which has been used to launch unmanned Proton Satellites, Salyut space stations, interplanetary probes to Mars, and Lunar-type vehicles to the moon.

For further details see text.

I. The RNV (Rakyeta Nosityel Vostok), the lauching rocket for one-man Vostok spaceships.

II. (*Above*) The spaceship Vostok used for the first manned spaceflight by Yuri Gagarin. The cylindrical section forms the last stage of the launching rocket.

(*Below*) April 12, 1961: Yuri Gagarin and his stand-in Herman Titov on their way to the rocket.

The RNV was not shown in the West until ten years after it was first used, namely at the 1967 Paris Air Show. The white-painted missile, 38 metres long, with the letters CCCP (= USSR) painted on its nose in bright red, and equipped with 32 gold-coloured exhausts, made a deep impression on the visitors—with the exception of the experts. Up to that moment practically everybody in the West had thought that the Russians owed their leading position in space travel to super-engines and possibly to super fuels. Now it became evident that the Soviets had in fact used one, not very powerful, but certainly reliable engine, several of which had been combined to produce the necessary capacity. The RNV (including all engines and smaller booster rockets) developed a thrust of 570 tons. Later versions were somewhat more powerful. The chief variants were the RNV-2 (Voskhod launching rocket, 650 ton thrust, 3 stages) and the RNS (*Rakyeta Nosityel Soyuz*, 700 ton thrust, 3 stages). Rockets based on the RNV were and are being used also for launching most of the Lunas, the Veneras, the Molnya communication satellites and some of the Kosmos satellites. It is thus an extremely reliable standard rocket, manufactured in large numbers and therefore relatively cheap.

There are several versions of the RD-107 also, including the RD-214, which is used in the first stage of the two-stage Kosmos launching rocket and is the third stage of the RNS. The Kosmos launching rocket is used for sending up light-weight scientific satellites (weighing a few hundred kilograms) within the Kosmos programme. The RD-214 is, in fact, an RD-107 with shortened exhausts and without the steering rockets which are characteristic of the RD-107. The "214" has a thrust of 78 tons. The second stage of the RNK (*Rakyeta Nosityel Kosmos*) is propelled by the RD-119, a simple engine with a thrust of only 11 tons. The RNK is another example of a standardised, and therefore relatively cheap, launching rocket, quite equal to the type of operations for which it is used.

It would be wrong to create the impression that the Russians restricted themselves to the development of rocket engines of relatively low capacity, but certainly those were the engines which were chiefly, and successfully, used in their standard programme. However, it was obvious that for space reconnaissance on a larger scale (space stations, flights to the moon and to the planets) much more powerful engines would have

Fig. 6. Soyuz launching rocket RNS (*Rakyeta Nosityel Soyuz*).

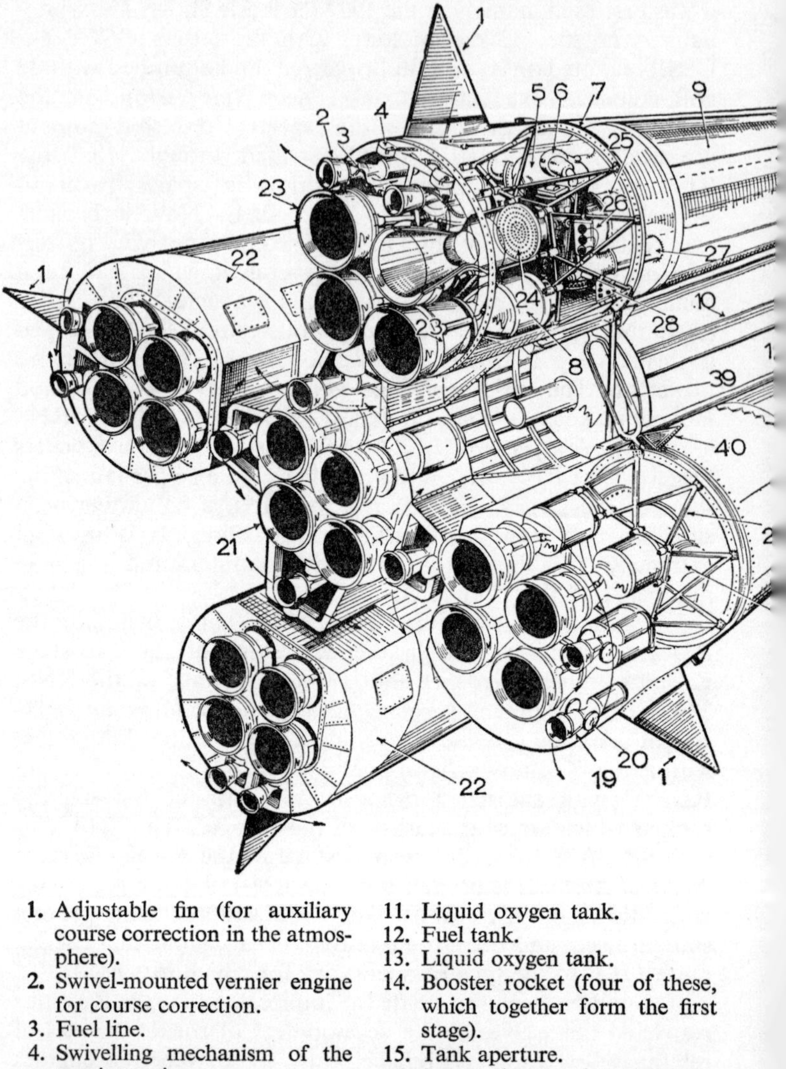

1. Adjustable fin (for auxiliary course correction in the atmosphere).
2. Swivel-mounted vernier engine for course correction.
3. Fuel line.
4. Swivelling mechanism of the vernier engine.
5. Heat-exchanger.
6. Gas-generator.
7. Frame for engine mounting.
8. Combustion chamber of RD-107 engine.
9. Fuel tank (kerosene).
10. Pipeline for liquid oxygen.
11. Liquid oxygen tank.
12. Fuel tank.
13. Liquid oxygen tank.
14. Booster rocket (four of these, which together form the first stage).
15. Tank aperture.
16. Separation point between booster rockets and second stage.
17. Booster aerial.
18. Turbopump.
19. Vernier engine.
20. Swivelling mechanism of vernier engine.

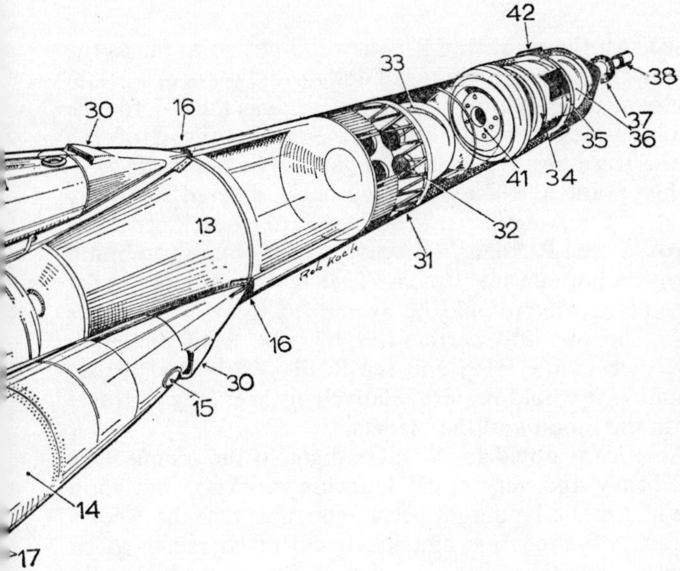

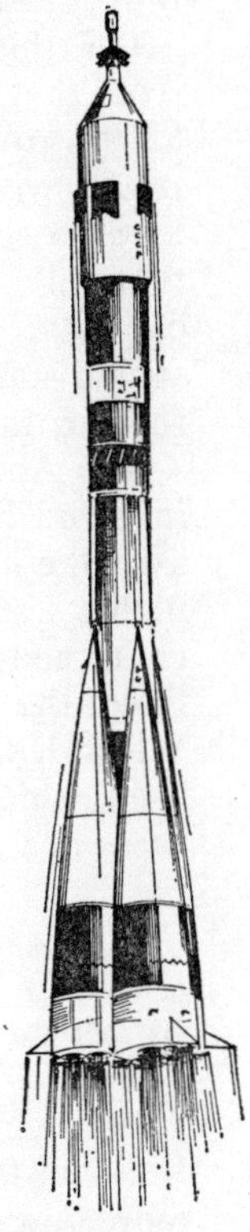

21. Exhaust RD-108 engine (second, central, stage).
22. Heatproof cover.
23. Circulatory fuel lines for cooling engine.
24. Fuel injection jets.
25. Liquid oxygen lines.
26. Electrical control system.
27. Fuel lines.
28. Regulators.
29. RD-107 engine mounting.
30. Ventilation ducts.
31. Third-stage mounting.
32. Four third-stage engines.
33. Third-stage fuel tank.
34. Propulsion element for Soyuz spaceship.
35. Command module Soyuz spaceship.
36. Cabin space of Soyuz (can also be used as airlock).
37. Escape system.
38. Separation rockets for escape system.
39. Cruciform booster link.
40. Streamlined cover.
41. Rear of Soyuz with engines for course correction and braking.
42. Stabilisation flaps for escape system.

to be used. All the same, the Russians did not go as far as the Americans who invested billions of dollars and years of work in the development of mammoth engines, such as the F-1 for the Saturn moon-rocket, which alone developed a thrust of 750 tons. The Saturn has five of these engines in its first stage.

On this point the Russian approach differed radically from that of the Americans. While the latter developed their Saturn rocket, the Russians' objective was to obtain maximum results with minimum investment. That is why they opted for a space station which could be assembled from components that might be put into earth orbit by rockets of moderate capacity (such as the RNS and the RNP—see below). From such a station it would require relatively little energy to travel further, to the moon and the planets.

The Americans opted for a direct flight to the moon. This involved heavy and very costly launching rockets, and soon after the first moon landing it became obvious that the Apollo flights were too expensive and incidental in character to be continued indefinitely. Too incidental: for on each Apollo flight the valuable rocket and all its equipment were lost. When departing from a space station, it is not only the station itself which remains, but also the moon rocket, which can carry out a shuttle-service between the station and the moon.

Super rocket

For launching their 12·2 ton scientific Proton-1 satellite, the Russians for the first time used a rocket which was considerably more powerful than the RNV and the RNS. On later occasions this rocket, the RNP (*Rakyeta Nosityel Proton*) was even used for launching 17·7 ton versions of the Proton. The RNP has a three times greater thrust than the RNV: its various stages together develop about 1800 tons. This rocket therefore comes between the Saturn-1 (total thrust 850 tons) and the Saturn-5 (total thrust 4350 tons). The capacity of the RNP is equivalent to approximately 60 million horsepower. The diameter of its upper stage is approximately 4·5 metres, as compared with 3 metres in the case of the RNV. Leonov, the first walker in space, assured me in Moscow that the RNP could be used for launching components for a manned space station. The rocket was, in fact, used for putting the Salyut in orbit. This was a 4-stage version of the RNP, developing a thrust of over 2000 tons. The RNP (see Fig. 7, p. 70) has six strap-on boosters.

There have often been rumours in the west that the Russians had developed an even more powerful rocket than the RNP. There were no means of checking these rumours. It is, however, certain that none of the known Russian artificial satellites, spaceships and space probes are so heavy that they would have required such a powerful rocket—more powerful even than the American Saturn-5.

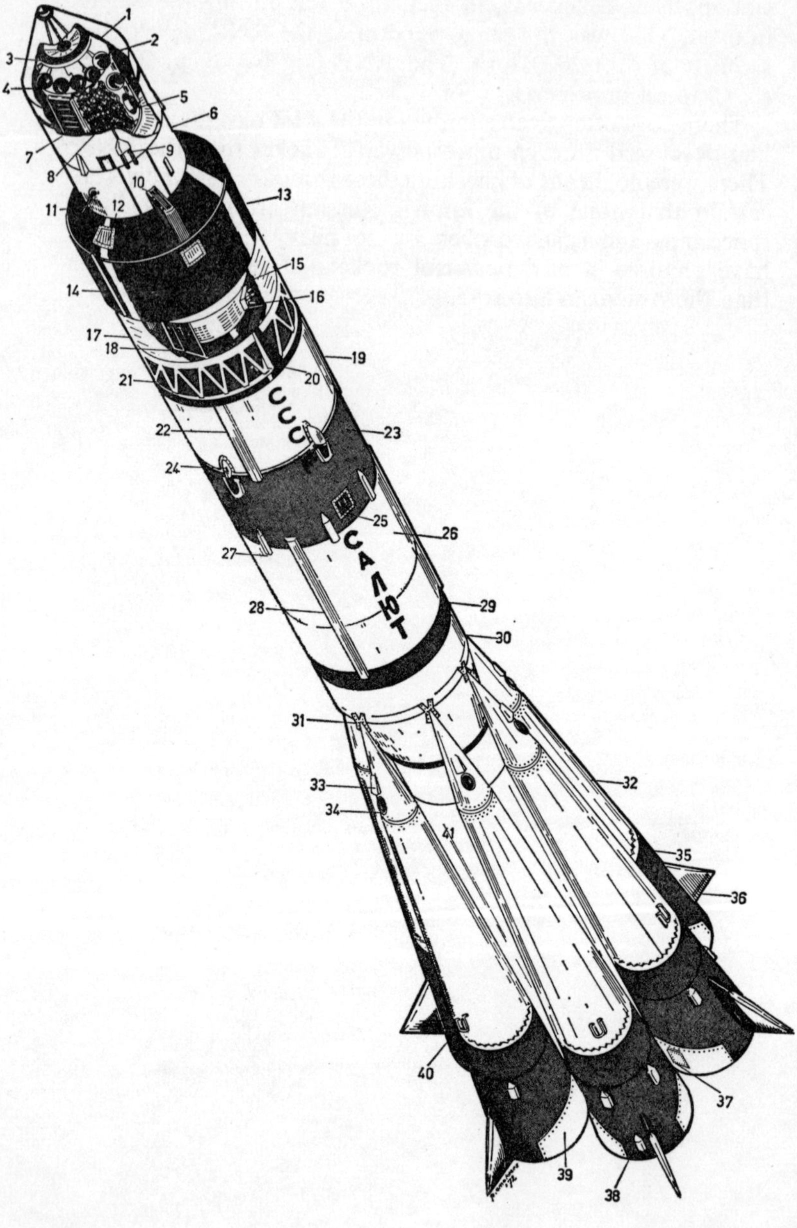
САЛЮТ
1
2
3
4
5
6
7
8
9
10
11
12
13
14
15
16
17
18
19
20
21
22
23
24
25
26
27
28
29
30
31
32
33
34
35
36
37
38
39
40
41

Fig. 7. RNP (Rakyeta Nosityel Proton)—Salyut.

1. Nose cone (protects Salyut during flight through atmosphere).
2. Transfer compartment of Salyut.
3. Coupling aperture of Salyut, with equipment for securing coupling probe.
4. Pressure bottles containing gases for attitude control jets.
5. Astrophysical telescope "Orion".
6. Salyut work space; section containing control- and steering equipment.
7. Solar cell panels (folded). They are deployed in space and convert sunlight into electricity.
8. Insulating layer protecting apparatus sensitive to temperature changes.
9. Separation mechanism of nose cone. On leaving the atmosphere the nose cone is split in two, exposing front section of Salyut.
10. Protective cover over electrical connections, etc.
11. Antenna.
12. Hatch to apparatus.
13. Salyut work space; section containing, among other things, apparatus for observing earth and physical training equipment. (See also fig. 27.)
14. Detachable hatch covering apparatus for observing and photographing.
15. Hemispherical-shaped rear of work space.
16. Attitude control jets of engine section.
17. Connection between Salyut and RNP carrier rocket.
18. Solar cell panels attached to Salyut engine section (see also 7).
19. Three external cable conduits.
20. Electrical connections between Salyut and RNP carrier rocket.
21. Cut-away of support section between Salyut and RNP.
22. Cable conduit (see 19).
23. Fourth stage of carrier rocket. Puts Salyut into orbit.
24. Connections between propellant tanks and fourth stage engine section.
25. Hatch giving access to apparatus.
26. Third stage of carrier rocket.
27. Connections between third and fourth stages.
28. Three external cables.
29. Connection between third and second (central) stage.
30. "Hammer-shaped" upper part of central second stage. (The lower section is narrower to accommodate six boosters).
31. Pressure points boosters (six.)
32. Six boosters. Jettisoned when burnt out. The boosters forming the first stage ignite simultaneously with central second stage.
33. Aerodynamic cover of ventilation aperture.
34. Tank aperture.
35. Booster engine section (including seven engine units of central stage).
36. Adjustable fin for additional course correction in the atmosphere.
37. Hatch to engine section.
38. Aerodynamic cover of starting equipment connections.
39. Heat-resistant cover of engine section.
40. Booster antenna.
41. Central section of second stage.

ROBOTS IN SPACE

Sputnik-1: the great surprise – Laika – Kosmos: jack-of-all-trades – Weather satellites – Cosmic spies – Space bombs – Radiation Satellites – Molniya solves communication problem

The great surprise

Time: October 4, 1957, 6 a.m. Moscow time. Place: Baikonur in the lonely Kazakhstan steppe, 250 km north-east of the Sea of Aral.

The white rocket gleams in the early sunlight. As yet it is imprisoned in the steel tentacles of assembly tower and cables. The dozens of technicians who, a moment ago, were scuttling to and fro like ants around the slender rocket, have retired into the blockhouse. The loudspeakers broadcast brief announcements, intended only for the handful of experts who are privileged to attend the launching. So far the world has no inkling of what is about to happen.

"*Gotovnosty dyesyat minut*" (Ten minutes to go).

In the underground launching centre a clock ticks away the last minutes. Sixty seconds to go. With a hissing sound the oxygen supply lines recoil; the electrical connections are abruptly disconnected. Now the rocket is dependent on its own sources of energy.

"*Zazhiganiye*" (ignition).

The four assembly towers still imprisoning the rocket as with giant hands, are slowly pulled away. The earth vibrates with the thunder of 32 rocket exhausts. Roaring, the rocket rises above the platform, faster and faster. The era of space travel has begun: Sputnik-1 is racing towards the cosmos. A few minutes later it is in orbit round the earth, freed from the carrier rocket and moving at a speed of 28 000 km per hour.

Mankind's first messenger in space is spherical—the classical shape of the moon. It has a diameter of 58 cm and weighs 83·6 kgs. Its four antennae, varying in length between 2·4 and 2·9 metres, inquisitively poke into space, tentacles of the

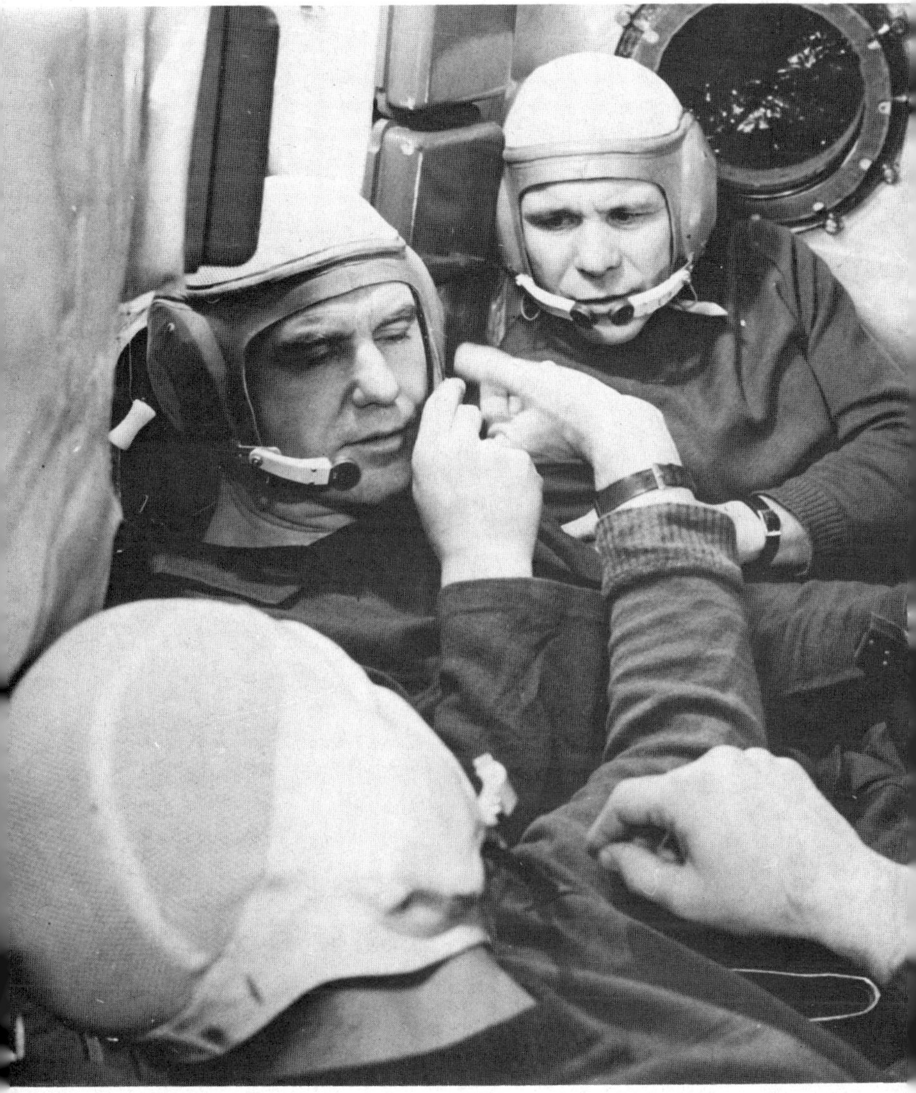

13. Alexei Yeliseyev (nearest the camera), Vladimir Shatalov and Yevgeni Khrunov in the cabin of Soyuz-4.

14. Boris Volynov, commander of Soyuz-5, ready for a parachute jump during his training programme.

15. (*Above*) The first cosmonauts to change capsule in space, Yevgeni Khrunov (left) and Alexei Yeliseyev (right), posing with Valeri Bykovsky.

(*Below*) With the docking of Soyuz-4 and -5 on January 16, 1969, the first experimental space station was established.

16. (*Above*) Soyuz-4 and Soyuz-5 linked together in orbit round the earth.

(*Below*) Cosmonaut Valeri Kubasov in the cabin of Soyuz-6. He was the first cosmonaut to carry out welding experiments in space.

scientific instruments and the two radio transmitters. The Sputnik's trajectory is sharply elliptical; its perigee is 228 km, its apogee 947 km;* and the angle of its orbit with the equator is 65 degrees. The artificial moon transmits bleeps on two wavelengths: 20·005 and 40·010 megahertz, that is to say approximately 15 and 7·5 metres. Any radio ham can easily receive on these wavelengths. Each signal lasts 0·3 seconds and is followed by a pause of equal length. The signals alternate on the two frequencies.

It is not until Friday evening that Radio Moscow announces the news. In Washington a cocktail party is going on at that time for delegates attending an international conference on the 1957–58 International Geophysical Year. Among the guests at the party are several Soviet scientists. Dr. Lloyd Berker, an American delegate at the conference, warmly congratulates the beaming Russians. Professor Blagonravov apologises for not having been able to tell the congress in advance that the satellite was to be launched. Significantly he adds: "It is the general principle of the Soviet Union not to announce anything before it is absolutely certain." He promises that the information which the experiment will undoubtedly yield, will be placed at the disposal of foreign scientists. Someone asks him for further details of the artificial moon. Blagonravov pretends ignorance. He says that he did not know that the satellite would be launched so soon. "Apparently the Russian scientists and technicians have been able to speed matters up and to launch the rocket sooner than expected. When we left the Soviet Union it was still being debated just which type of satellite was to be used." Blagonravov said he did not know either from where the satellite had been launched. "There were three possible launching bases. I don't know which one was chosen." Later it became known that it had been Baikonur.

Considered after the event, it seems almost incredible that Sputnik-1 came as a surprise to the West. In previous years Soviet scientists had often hinted, and sometimes declared openly, that the Russians would launch a satellite on the occasion of the International Geophysical Year, which had as its objective a world-wide study of the earth and its natural resources. But little attention had been paid to the statements of "those backward Russians".

*Perigee: Point of trajectory nearest to earth; apogee: point of trajectory furthest from earth.

To begin with, Col. Gregory A. Tokayev, the Russian rocket expert who in 1948 had sought refuge in England, had in that same year more than once expressed the conviction that his former fellow countrymen intended to put a satellite into orbit round the earth. On the Soviet side, however, nothing was said on this subject until 1953. Stalin had demanded that absolute secrecy should be maintained on the subject of rocket development in the USSR. But hardly had the dictator been laid to rest in the Mausoleum when Professor Alexander Nesmeyanov, then president of the Academy of Sciences, declared on November 27 of that year at a meeting of the World Peace Council in Vienna: "Science has now reached a stage where it makes sense to speak of sending a rocket to the moon and of creating an artificial earth satellite."

In May 1954, moreover, a warning was sounded in America itself, the country which only a few years later would be so dumbfounded by the launching of Sputnik-1. On the occasion of the third Symposium on Space Travel in the Hayden Planetarium in New York, George P. Sutton of North American Aviation* said that the Russians would soon be capable of launching both manned and unmanned satellites. His remark that from such a satellite the Russians would "bombard" the earth with radio propaganda and might even threaten peace-loving nations with weapons from space, was typical of the international political atmosphere of that time. . . .

On July 2, 1955, President Eisenhower announced that the United States would launch a satellite on the occasion of the International Geophysical Year. At that time no "reasonable" person doubted for a moment that the United States would be first. As a matter of fact there were singularly few experts who realised that there was any question of scientific and technological competition. Yet this might have been suspected when two days later Professor Leonid Sedov, who at that time had been head of the Interdepartmental Commission for Interplanetary Communications for just over a year, announced cautiously that the Russians were planning an earth satellite. Sedov made his statement in the Russian Embassy in Copenhagen during the sixth congress of the International Astronautical Federation.

*This company subsequently built the Apollo spaceship.

"Lately a great deal of attention has been devoted in the USSR to investigations in the field of interplanetary communications, particularly as regards the creation of an artificial earth satellite. In my opinion it will be possible within two years to launch such an artificial satellite, and we already possess the technical ability to launch satellites of various sizes and weights." The last sentence was obviously meant to indicate the fact that the Russians would be able to launch much heavier satellites than those planned within the framework of the American Vanguard project, for the latter weighed barely three pounds.

In the course of the following year, 1956, there were more and more indications that the Russians were in earnest with their preparations for launching "an artificial moon", but the West continued to give little credence to the Russian pronouncements. To many people it was difficult to accept the possibility that a country which up to 1917 consisted largely of illiterates, would be capable of such a technical achievement.

On September 11, 1956, a conference took place in Barcelona, in preparation for the International Geophysical Year. At this conference Ivan Bardin, a member of the Academy, announced officially that in the course of the IGY his country would launch a satellite with the aid of which it "would be possible to measure air pressure and temperature, cosmic rays, micrometeorites and the magnetic field of the earth."

In June 1957, Professor Nesmeyanov announced in the Soviet press that the Russians would launch a satellite into space within the next few months. In May, June, July and August, Soviet radio amateurs were asked to listen for an artificial moon which would broadcast on wavelengths of 7·5 and 15 metres. They were even given instructions on how to build a suitable receiver for this purpose. In early October the Russians were making the final preparations in conditions of the greatest secrecy . . . although as yet there was no launching rocket on the launching base. Unlike the Americans, the Russians check both rocket and satellite in the assembly hall, where it is placed on a trailer running on rails. It is here, in this hall, that the countdown begins, weeks in advance. Once the rocket leaves the assembly hall, it is, in principle, merely a matter of placing it in an upright position and firing it. If all goes well, the Russians can fire a different rocket—manned or unmanned—from the same launching pad every day. This is of course a very economical method, for a large part of the money invested in spaceflight is spent on the ground installations. The fewer installations that have to be built the better.

This aspect of the Russian launching method is hardly realised in the West, and as a result it caused amazement that the Russians were able to launch, for instance, Soyuz-6, -7, and -8 in such rapid succession. Nevertheless it is probable that on this occasion two installations were used.

October 2. . . . The rocket is practically ready. The satellite has been mounted, the nose-cone is attached. Once again Chief Designer Sergei Korolyov and his men go over every point of the check-list.

October 3. Dozens of engineers and workmen are gathered in the assembly building. Members of the Government Commission responsible for the execution of all future spaceflights are also present. The enormous doors open and the transporter with its load is trundled slowly and carefully outside. It is not far to the launching base which is ready to grasp the rocket in its metal tentacles.

Hydraulic cylinders bring the 30-metre colossus to an upright position. The shape of the rocket is by now well known: a central main stage surrounded by four large booster engines. On top of the main stage is the nose-cone which conceals the spherical satellite. On later occasions the same rocket type is to be used for manned spaceflights and for flights to the moon and the planets, but then there will be an additional stage. . . .

As soon as the rocket is upright, fuel and liquid oxygen are pumped in. A few small technical faults are dealt with. The countdown runs its course practically without a hitch.

And then, on October 4, 1957, at 6 o'clock Moscow time: "*Zazhiganiye!*" The era of space travel has begun.

Reactions in the West varied between amazement and barely concealed envy. Clarence B. Randall, an adviser to President Eisenhower, thought fit to remark that the new Sputnik was only a "pathetic bauble", while someone else declared: "Anyone can fling a bit of metal into space."

But most reactions—though on the whole interpreting the Americans' disappointment—were purely appreciative.

Of course it was annoying that the United States, which, with the necessary self-confidence, had announced the launching of a satellite as early as 1955, was still not ready for it. The fact was that the Navy were having trouble with the Vanguard rocket which had been specially developed for the purpose. On December 6, 1957, that is, two months after the start of Sputnik-1, the first Vanguard collapsed on the platform and exploded after it had been fired. On February 5, 1958 a second rocket lifted off, but exploded after reaching 7 km. Only the third rocket of this type succeeded in putting a miniature satellite (jokingly called "a grapefruit") into orbit round the earth. This was, however, not the first American satellite. Immediately after Sputnik-1, Wernher von Braun, who was working for the American Army, had been instructed to launch "an artificial moon" with the means at his disposal. This was Explorer-1, which circled the earth on January 31, 1958.

Naturally the Russians took each and every opportunity to declare that once again the socialist system had been proved superior to the capitalist one. (What, one wonders, can the Soviets have thought when in 1969 the Americans were the first to reach the moon?) The well-known Russian clown Karandash (Pencil) had his own individual reaction to the American flops. He entered the circus arena carrying a small balloon. The balloon exploded and his "feed" asked: "What was that?" "That was a sputnik," replied the clown. For a moment the audience gasped, until Karandash smiled and said: "The American sputnik, of course!"

Excitement about Laika

The fact that the Russians had had a scoop was, of course, in itself not all that important. Nevertheless the Americans fully realised that they were considerably behind in the field of rocket propulsion. Rockets, after all, could be used for conveying A- and H-bombs as well.

The launching of Sputnik-2 on November 4, 1957, that is, only a month after the start of number one, emphasised the gap even more, for this second sputnik weighed no less than half a ton. Some mathematicians rapidly calculated that this meant that the launching rocket must have had an initial thrust of around 500 tons, and subsequently it became apparent that they were not far out. And the Vanguard produced only fourteen tons! Even America's first inter-continental rocket, Atlas, only reached a 180 ton thrust!

As if the launching of Sputnik-1 was not sensational enough, the start of the second sputnik created a storm of excitement all over the world, for the satellite carried a living creature, the bitch Laika. The dog, which lived for ten days in its miniature spacecraft, proved for the first time that the organism of a vertebrate (including that of a human being) is able to stand the conditions of a prolonged spaceflight, and in particular that of weightlessness. Although at the time the Russians left the matter open, even hinting that the animal might be returned to earth, they never intended to bring her back. Sputnik-2 could not survive re-entry into the atmosphere, for it had neither heat-shield nor retro-rockets. When the Soviets launched their Laika they knew that the animal would not come out of the experiment alive. There was sufficient food and oxygen in the narrow compartment for only ten days. Apart from the cabin which housed the dog, Sputnik-2 contained a spherical section for instruments and radio transmitters, exactly as in Sputnik-1.

After having remained out in space for ten days without any noticeable physical effects, Laika died of lack of oxygen. Sputnik-2 completed a total of 2370 orbits, after which it was destroyed on re-entry into the atmosphere during the night of April 13 to 14, 1958.*

The third Sputnik

A less spectacular, though no less important event was the launching of Sputnik-3, an artificial satellite with a weight of 1327 kg—almost incredible at that stage. For this operation too, Russia's "jack-of-all-trades", the RNV rocket, was used. The satellite did not carry a living creature, but was filled to the brim with a varied collection of ingenious measuring apparatus, designed to provide as much information as possible concerning the immediate environment of the earth at any par-

*"Cosmic Biology" (p. 95) deals with the bio-medical aspects of the flight.

ticular moment. Among other things, Sputnik-3 was to send out details about the earth's magnetic field, about micrometeorites, cosmic rays, positively charged atomic particles and pressure inside and outside the satellite. The cone-shaped satellite was 3·57 m long. For the first time solar energy was used for the instruments and the transmitters. They lasted for more than two years, which proved that in cosmic space, where there is plenty of sunlight, solar cells have many advantages over conventional batteries. To begin with, the latter are so heavy that only a few can be sent up, and consequently the satellite cannot rely on their power for long.

Apart from the solar cells, Sputnik-3, which initially orbited at between 226 and 1881 km, contained another important innovation, namely a unique temperature regulation system which subsequently was to be used on a large scale in manned Russian spacecraft. Maintaining a constant internal temperature is essential not only for manned spaceships, but also for unmanned satellites. Some instruments function best in an even temperature, and it is therefore important to maintain a constant level.

What is the temperature in space? In the early years of space travel it was thought that space would be "intensely cold". Space travellers would have to be protected from this cold. However, if we consider what temperature means, we realise that this cannot be true. Environmental temperature—which is what matters in this case—is the average kinetic energy of air particles. Since in space these particles are virtually lacking, there can be no question of temperature. Space has no temperature, since—at least in initial approximation—it contains "nothing". Naturally a body in space can have a temperature of its own, for solar energy can be transferred to it by means of radiation. Now the question is whether this body can lose this heat-energy quickly enough. In other words: the temperature of a body in space is determined by the relationship between the quantity of heat absorbed and radiated. The whiter the body, the less heat it will absorb—hence the preference for white spacesuits, particularly for space- and moon-walkers, who are exposed to direct solar radiation. To regulate the temperature of a spaceship and prevent the outside being overheated, it is possible either to revolve the capsule, or to use some sort of sun-blind by which the reflection-coefficient of the surface may be varied. The third of the sputniks had 16 such louvre blinds, which were adjusted by a mechanism reacting to the temperature inside.

Owing to the high orbit in which it had been launched, Sputnik-3 had a very long life: 691 days. It burned up on April 6, 1960 on re-entry into the atmosphere. In the course of its cosmic existence it had circled the earth 10,037 times and transmitted an enormous volume of information to its earthly masters. As a result it was possible to make a serious start on the preparations for a manned spaceflight, which was to materialise just over a year later. All subsequent Sputniks were direct precursors of the manned Vostok spacecraft.*

Kosmos: a blanket name

The most remarkable and most mysterious programme of unmanned Russian satellites is undoubtedly the Kosmos-series. Hardly a week passes by without the newspapers publishing the brief announcement that the Soviets have launched the umpteenth "Kosmos". The Tass communiqués usually excel in their stereotyped choice of words. Generally the announcement ends with the triumphant: "The scientific apparatus on board is functioning normally. Information received is being processed by the co-ordination and calculation centre." Moreover, Tass always faithfully mentions that the new sputnik's objective is further research of space, in accordance with the programme announced on March 16, 1962. This programme, which was published on the occasion of the start of Kosmos-1, read as follows:

> "The scientific programme includes the study of the following subjects:
> —the concentration of charged particles in the ionosphere for the purpose of investigating the distribution of radio waves;
> —corpuscular flows and low-energy particles;
> —the energy composition of the Earth's radiation belts, for the purpose of further evaluating the radiation danger in prolonged spaceflights;

*See "Cosmic Biology", p. 95.

III. (*Above*) Interior of the Vostok. Above the cosmonaut's head one can see the "Vzor" optical apparatus.

(*Below*) Pavel Belyayev and Alexei Leonov (right) before their take-off in Voskhod-2.

IV. *(Above) Left* Alexei Leonov during his space walk. *Right* The earth, as seen by Zond-7 on August 8, 1969 from a height of 70,000 kilometres. (*Below)* Luna-9, the first capsule to make a soft-landing on the moon and which later relayed photographs back to earth.

—primary cosmic rays and the variation in their intensity;
—the Earth's magnetic field;
—the short-wave radiation of the sun and other heavenly bodies;
—the upper layers of the atmosphere;
—the effects of meteoric matter on the construction of space vehicles;
—the distribution and formation of cloud patterns in the Earth's atmosphere.

Moreover many elements of space vehicle construction will be checked and improved.

So far the Tass communiqué. In practice it had become obvious that several points might have been added to this short list.

The Kosmos artificial satellites may roughly be divided into five categories:

—small scientific artificial satellites (weighing a few hundred kg);
—reconnaissance satellites ("cosmic spies");
—prototypes of manned spaceships;
—meteorological sputniks;
—"space bombs" and "satellite destroyers";
—moon- and planet-reconnaissance satellites which have failed to get out of earth orbit.

It cannot be doubted that nearly half the artificial satellites launched within the framework of the Kosmos programme serve solely scientific objectives. Representative of this category is Kosmos-97, which was put into an elliptical orbit on November 26, 1965, for the purpose of measuring radiation. It is known that the earth is surrounded by two belts of electrified particles moving at terrific speed. Originally it was thought that these would form a serious threat to the health of travellers to the moon. These belts, the so-called Van Allen Belts, are situated at around 3000 and 16 000 km respectively. Some scientists consider that it is rather a question of one belt—they merge into each other gradually. The belts follow the lines of force of the earth's magnetic field and this is hardly surprising, for they consist of negatively and positively charged atomic particles: electrons and protons. Radiation intensity within the belts shows considerable variation, linked among other things with the radiation intensity of the sun. The weaker offshoots of the belts were traversed by another Kosmos satellite, Kosmos-110, carrying the dogs Ugolyok and Vyetyerok.* Another scientific artificial satellite, also launched to measure the intensity of cosmic rays, was Kosmos-145. Various other Kosmos satellites have been specially designed to discover the secrets of the earth and its atmosphere. The 500th Kosmos was launched on July 10, 1972.

Meteo-Kosmos flights

On June 25, 1966, President de Gaulle of France was the first western guest at Baikonur, where he witnessed the start of Kosmos-122, the first of the Russian series of meteorological artificial satellites. This, as yet experimental, meteorological satellite was equipped with television cameras and infra-red sensors in order to photograph the clouds by day and by night. In 1967 it was followed by a pair of Kosmos satellites supplementing each other, numbers 144 and 156, which were both launched into polar orbit and transmitted masses of interesting information which added a new dimension to Soviet meteorology. (Weather satellites such as Tiros, Nimbus and Essa had already been available to the Americans for some time.)

Nowadays operational Russian weather satellites are launched under the name of "Meteor". New satellites are continually being launched to complete the network (each part of the earth must be photographed several times a day) and to replace those that have dropped out.

Cosmic spies

As a rule weather satellites photograph fairly large areas at a time, thus providing a clear idea of the existing situation. Kosmos-122 alone, for instance, each hour recorded 30 000 square km of the earth's surface on film. Of course it is also possible to photograph very small areas from a satellite; this brings to light some most interesting details. In other

*See "Cosmic Biology", p. 95.

words: a satellite makes an ideal spy: practically invulnerable (unlike the American U-2 which was brought down over Russia in 1960), extremely reliable (it cannot desert) and highly productive. Needless to say both the Americans and the Russians foresaw this possibility years ago. Both nations therefore industriously launch artificial satellites with the regularity of clockwork, satellites which use up miles of film and then return to earth, where they are recovered by experienced teams, specially trained for this kind of work. The American spy satellites (more politely called "reconnaissance" satellites), which usually start from the Californian bases Point Arguello and Vandenberg, are not even given names, merely numbers. The Soviets hide their cosmic espionage activities under the cloak of the Kosmos series.

Approximately 50 per cent of the Soviet artificial satellites belong to this category. They orbit the earth for either eight or twelve days at a stretch at between about 200 and 300 km, and then land in the deserted steppes of Central Asia. According to some western sources of information, they are mainly unmanned versions of the Vostok.

Although the Russian press does mention the launching of these satellites, it only publishes details about their trajectories: height, duration and angle with the equator. The landing is never announced, but thanks to American and English tracking stations, which every two weeks publish a highly detailed report, we know exactly when a Russian scout has returned to its socialist fatherland. These American and English observations are exceptionally accurate. In one case, when a Soviet artificial satellite had exploded shortly after going into orbit, the exact trajectory of each fragment—more than 100 in all—was published!

Initially reconnaissance satellites were launched from Baikonur, but later the new Plesetsk base, in the neighbourhood of Archangel, was increasingly used. Like many weather satellites, the reconnaissance satellite thus goes into polar orbit and in the course of each 24 hours it can survey the entire earth as it revolves below.

In addition the Russians use their "Kosmos" denomination fairly intensively for testing new types of spaceships which are later to be manned. Thus, on October 6, 1964, they launched Kosmos-47, which went into orbit at between 177 and 413 km and returned after exactly 24 hours. Six days later Voskhod-1, with Komarov, Feokistov and Yegorov on board, was launched. Its trajectory was practically identical to that of its unmanned predecessor, and this spacecraft also returned after 24 hours. Unmanned Soyuz spaceships were Kosmos-133, -140, -146, -154, -186, -188, -212, -213 and -238. With numbers 186 and 188 the Russians for the first time achieved a link-up in space. Later they repeated the operation with Kosmos-212 and -213. Both these experiments were in preparation for the docking of Soyuz-4 and -5 in January 1968. Kosmos-154 was the final dress rehearsal for the start of Soyuz-1, which, after a flight of just over 24 hours, crashed as the result of the non-functioning of a parachute, killing Cosmonaut Vladimir Komarov.

Space bombs

On November 7, 1965, Tass issued a report on the military parade which had taken place in Red Square that day on the occasion of the 48th anniversary of the revolution. "The column of rocket troops ended with orbital rockets with atomic warheads, which are capable of hitting any aggressor unexpectedly, after making one or more orbits round the earth."

In America the system to which this report referred is called "FOBS", an abbreviation of Fractional Orbit Bombardment System. It consists of putting a satellite containing a hydrogen bomb into orbit and making it "land" on the enemy target. To do so, the retro-rockets must be fired approximately 800 km before reaching the target area. This Russian "terrorist weapon", as George C. Wilson called it in the *Washington Post*, renders the radar warning systems on America's northern frontier and western seaboard useless up to a point. Such a FOBS artificial satellite may be said to "come in by the backdoor", as Khrushchev once boasted. In principle this system also makes it possible to leave a hydrogen bomb in orbit until the time is thought ripe to make it come down. Another "advantage" is that a FOBS takes about ten minutes less to reach its target than a "conventional" intercontinental nuclear warhead, which describes a large hyperbola, reaching a height of approximately 1300 km. These Russian satellites orbit at a height of 200 km only. It is hardly surprising that shortly after the news became known, the US government was urged to develop rockets capable of eliminating artificial satellites. The difficulty is that one cannot be certain whether the object is a FOBS satellite or not. This becomes reasonably certain only when it is observed that the retro-rockets have been fired and that the satellite is approaching. . . .

Naturally enough the Russians considered it necessary to test their system from time to time. They did so with Kosmos-139, -218, -298, -354 and -365, among others. All these artificial satellites had a trajectory at a height of between about 144 and 210 km and landed on an imaginary target somewhere in the Soviet Union after approximately one orbit. Since as a rule they did not complete even one orbit, the revolution time was never announced. Tass only mentioned perigee and apogee, and to experts this was reason enough to be suspicious. It is in any case unlikely that these "space bombs" contained a nuclear payload.

Of course space bombs may complete several orbits before they are brought down. In that case the system is called MOBS (Multiple Orbit Bombardment System). It seems that the Soviets have tested this system on a few occasions. In times of international tension they would be able to "place" a ring of MOBS-satellites around the earth. These space bombs could be brought down at any time and place to carry out their destructive task.

A number of American defence specialists (at the time including McNamara) are of the opinion that the FOBS- and MOBS-bombs would be far less accurate than bombs launched with "conventional" intercontinental rockets. All in all, however, the Americans regarded the Russians' space bombs as a sufficiently serious threat to develop a radar system which can see "beyond the horizon", in order to detect the low-flying space bomb at a relatively early stage. If at some future date nuclear weapons were really to be placed in space, it might be necessary for the "sender" as well as for the "addressee" to be able to destroy such a bomb in orbit. The Russians already seem to be aware of the fact that such a FOBS or MOBS might possibly not heed the master's command. According to some western experts they are therefore developing a system of destroyer-satellites, which will of course enable them to liquidate hostile satellites as well.

Elektron

For a long time the Russians were very pessimistic about the danger of radiation on distant voyages into space. In order to increase their knowledge, in particular about the composition of the Van Allen Belts, they launched their Elektron satellites. Electron-1 and -2 were launched on January 30, 1964 by one rocket. The first satellite was separated while the final rocket stage was still functioning, so that number two went into a considerably higher orbit. The same procedure was followed with Electron-3 and -4, launched on July 10, 1964. The trajectories of both these pairs had been carefully planned: they traversed both radiation belts, which are located respectively at around 3000 and 16 000 km from earth. The satellites measured not only the constantly changing intensity of the belts, but also the effect which the magnetic field of the earth and the particles sent out by the sun had on this intensity. It is now clear that the intensity of the belts varies with the electric particles "pumped" into them by the sun. In this connection it is of the greatest importance that the Russians launched their Elektrons in a year when the sun was relatively inactive. As a result the measurements recorded were highly suitable for comparison with those obtained (by considerably more modest means) in the course of the International Geophysical Year (1957–58). The radiation activities of the sun show a cycle of 11 years.

Prognoz

A special satellite for research into solar activity and its effect on the earth was launched on April 14, 1972. This satellite, the "Prognoz", which weighed only 845 kg, was put into an extended elliptical orbit: between 950 and 200 000 km from earth, so that it penetrates far into space. Prognoz orbits the earth once every 97 hours.

The main purpose of this satellite, which will undoubtedly have a very long life, is to obtain greater insight into the gamma-particle and x-ray radiation of the sun, which greatly influences the weather and communications and which, moreover, might have an adverse effect on the health of cosmonauts far out in space. Prognoz is intended to contribute to the possibility of making reliable prognoses concerning solar radiation in the future.

Proton

A "mighty new rocket", as Tass put it, was used for launching the super-heavy satellites in the Proton series, which put Russia in the lead for some time. The Protons, of which the first was launched in 1965 and the fourth in 1968, had weights varying between 12·2 and as much as 17 tons. They carried very heavy instrumentation for measuring super-charged cosmic particles, chiefly protons. Protons are the nuclei of hydrogen atoms, the most dangerous elements of cosmic radiation.

The Protons concentrated mainly on exploring those sections of space surrounding the earth where manned satellites operate, that is, between 190 and 630 km. Mystery still surrounds a sub-satellite (listed in the catalogues as "1965—64C"), carried by Proton-1. After exactly eight days it made a (controlled?) landing near Magnitogorsk, 500 km from the Sea of Aral. Possibly—but this is sheer conjecture—it was an unmanned version of the Soyuz capsule. The first obvious test with such a capsule was made when Kosmos-133 was launched on November 28, 1966.

Polyot

"Polyot" (flight) was the name given to two satellites which certainly did form part of the Soyuz project. The Polyots, which were launched on November 1, 1963 and April 12, 1964 respectively, were the first artificial satellites able to carry out their own course correction. Both the height of the trajectory and the angle with the equator were changed several times by means of the engines. Thus the foundations were laid for spaceships which could carry out approach- and docking operations: the Soyuz. The changes in the angle with the equator were necessarily very slight, as it is these corrections which require the most energy. To take an extreme case: a spaceship in equatorial orbit (inclination 0 degrees) is to change over to polar orbit (inclination 90 degrees). This means that first of all its forward thrust (speed: 28 000 km per hour) has to be cancelled out. Obviously this in itself requires a rocket of equal power to that which has put the spaceship into orbit. An equally powerful rocket is then needed to achieve the required speed after the 90 degree angle correction has been carried out. Of course this is an extreme example, but it does show how much energy is required for large adjustments in the angle with the equator. On the first day of its existence Polyot-1, which had been put into orbit at a height of 339 km, increased its apogee to 1437 km. Its inclination, however, was corrected only very slightly, namely 0·02 degrees.

Yantar

One of the most mysterious unmanned Russian space-shots was the launching of Yantar-1 (the word means "amber") in October 1966. Shortly after the start the usual rumours went the rounds. Actually there was no question of an orbital satellite: Yantar-1 was a very small rocket, equipped with an ion propulsion device. The apparatus was put up to a height of 400 km by a geophysical rocket. At that height the Yantar was disconnected and the ion propulsion device switched on. This operation provided insight into the action of ion propulsion devices, small rockets, in which very hot gas particles are accelerated by a magnetic field. Ion propulsion devices as yet provide very little thrust and are therefore not suitable for putting a spaceship into orbit. The gas particles, however, reach a very high exhaust speed (40 km per second in the case of Yantar-1, as compared with 2·5—3 km per second in a conventional rocket engine), and as they can go on working for very long periods, they could eventually give a spacecraft enormous speed.

As a matter of fact the Russians had already equipped their Zond-2 with an experimental ion propulsion device. Similar devices (which have a diameter of only a few centimetres!) were used in Voskhod-2. They were of course too weak to be used for course corrections, but nevertheless they did turn the spacecraft on its own axis. The Soyuz craft, too, carry small ion propulsion devices as additional equipment.

Molniya: a "television transmitter" 40,000 km high

On "Radio Day" in 1967, the Russian daily *Krasnaya Zvyesda* (*Red Star*) published on its front page a beautiful photograph of the new Moscow television transmitter. This slender structure, admirable from an architectural as well as a technical point of view, rises to a height of more than 500 metres over the Soviet capital. As usual, the *Red Star* paid tribute to "the workers" and engineers who had created this transmitter, the highest in the world.

But with all respect to the Russians, even this transmitter, in spite of its enormous height, has its limitations, for television signals do not follow the curvature of the earth any more than do rays of light. The transmitter can therefore not reach receivers below its horizon. True, for the Moscow transmitter this horizon is 200 km away, but the area thus covered is still minimal compared with the vast surface of this immense country. The problem of communications within the Soviet Union has existed for a long time and, after electrification of the country, it was probably the most important technical problem to occupy the attention of the government shortly after the revolution. As early as the 1930's, Russian engineers developed an ambitious plan for the exchange of radio messages between Moscow and Leningrad, just over 600 km away as the crow flies. The project consisted of a plane, equipped with relay apparatus, flying in a small circle halfway between the two cities. The apparatus in the plane was to receive, amplify and relay to Leningrad programmes broadcast in Moscow, and vice versa.

According to the originators of the idea, the same apparatus could be used for the exchange of messages. The plan was never carried out. For practical reasons the idea was shelved and instead a very costly cable was laid between the old and the new capital. For communication with more distant towns cables were also used more and more, although it remained impractical to provide an efficient link between places as far apart as Moscow and Vladivostok (over 6000 km). Thus, having long since solved the problem of electrification (even the smallest hamlet in the Soviet Union has a primitive electricity supply), the Russians still have communication problems. This applies to many other parts of the world. The Americans launched their Early Bird, Relay and Syncom satellites and the Russians finally found an excellent solution in their Molniya a communication satellite. The first Molniya* appeared in space on April 23, 1965 and in subsequent years the series increased rapidly.

When the Russians started their project for an artificial communication satellite in 1950—that is, no less than seven years before the start of Sputnik-1—they fully realised that there were several possibilities:

—A communication satellite in circular orbit a few hundred kilometres above the earth's surface. Such a satellite would, however, still only reach a comparatively limited area and would, moreover, completely circle the earth every 90 minutes. As a result of its constantly changing position, it could connect any two places for only short periods at a time. True, this problem could be circumvented by installing a data-storage system, but in order to be able to connect any two places permanently, it would be necessary to launch about fifty of these satellites.

—A "synchronous" satellite. Such a satellite moves in equatorial orbit at a height of 35 000 km. It circles the earth in exactly 24 hours (the higher the trajectory, the longer it takes), so that, although moving at a very fast speed, the satellite remains "suspended" over the same point on the earth's surface, since the earth itself revolves on its axis every 24 hours. In this way an imaginary television transmitter would be established, which would be high enough to survey nearly half the earth. Only three of these synchronous satellites (of which the American Intelsat is a good example) spaced 120 degrees apart in the same orbit, would be sufficient to guarantee permanent communication between all parts of the earth.

—A satellite in a markedly elliptical orbit, relatively low over the southern, and very high (40 000 km) over the northern hemi-

* "Molniya" is Russian for "lightning".

sphere. Such a satellite would have an orbiting time of 12 hours and would, for nearly ten hours at a time, reach the extreme east and west of the Soviet Union simultaneously. During the rest of its orbiting time it would pass over the United States.

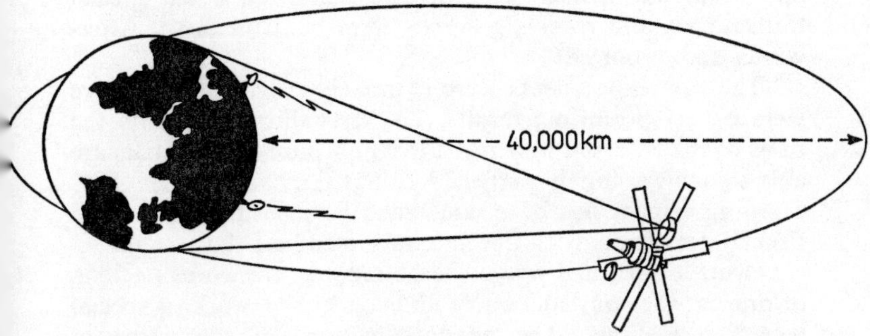

Fig. 8. Trajectory of a Molniya communications satellite (drawn approximately to scale). The satellite "surveys" a very large part of the northern hemisphere.

The Russian engineers finally chose the last alternative, for the following reasons. Obviously the first alternative was rejected, if only because it would be difficult as well as expensive to create and maintain a network of fifty communication satellites. The second alternative would be ideal, but geographically the Soviets are in a less favourable position than the Americans. Baikonur, the Russian launching base closest to the equator, is situated at a latitude of 48 degrees north and it requires an incredible amount of power, and several complicated manoeuvres, to put a satellite into equatorial orbit from this point. The Russians therefore opted for the only remaining alternative: a satellite in a markedly elliptical orbit. Thus project Molniya was born.

The installations on board the Molniya satellites are suitable not only for relaying television programmes—although this is their most spectacular achievement—but can also be used for relaying radio transmissions, telephone conversations and telegrams.

The Molniya satellites are moreover equipped with cameras for meteorological photography. On instructions from earth different filters can be used. These photographs form a welcome addition to the material obtained by means of special meteorological satellites.

Immediately after Molniya-1A had been put into orbit, communication experiments were carried out between Moscow and Vladivostok, and Moscow and Paris, for which special transmitting and receiving aerials were built on earth. Afterwards Tass reported:

"The first experiments were extremely successful and have yielded very promising results. The tests showed that, on the basis of the SECAM system,* Soviet and French scientists are able to achieve highly perfected colour television.

An agreement has been concluded between the Soviet and French governments to develop and introduce the system."

Once Molniya had proved to be reliable, there was nothing to prevent the construction of an extensive network of special reception stations. The network was given the name of "Orbita". Within less than a year more than twenty "Orbita" receiving stations, each equipped with a large parabolic aerial, had been constructed in the most remote areas of Siberia, in the icy north, the far east, and southwards towards the Caucasus and Persia. Several Molniyas at a time are constantly in use. New satellites regularly replace the older ones.

Whereas formerly millions of Russians in the vast Siberian steppes could only see programmes from Moscow which had been recorded on film several days before, their television sets are now linked directly to the capital via space. This, of course, is especially important where news bulletins and other topical programmes are concerned.

And although the pictures have travelled thousands of kilometres through space, their quality is exceptionally good. The West saw proof of this towards the end of 1967 when, on the occasion of the fiftieth anniversary of the October Revolution, a programme was screened entitled "An hour in the life of the people". In this live transmission the various Molniya satellites in use at that time played a decisive part.

But the Molniyas not only relay television programmes. Several dozen channels in the ultra shortwave band have been reserved for telephonic and telegraphic communications

*SECAM-system: French system for transmitting television pictures, slightly different from the PAL-system used elsewhere in Europe, and the NTSC-system applied in the United States.

between widely separated points in the Soviet Union. In this connection Lyustibyerg, scientific commentator of the Novosti News Agency, wrote: "This is equivalent to laying thousands

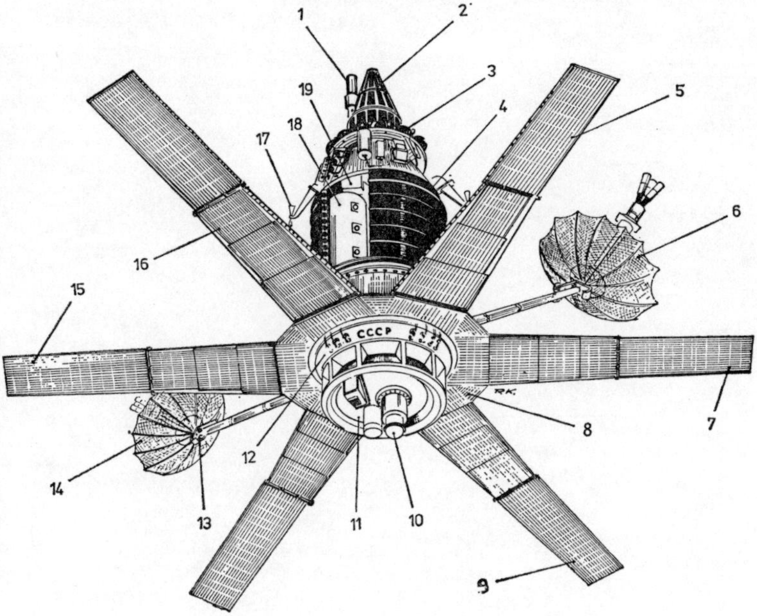

Fig. 9. Molniya-1 communications satellite.

1. Earth directional device.
2. Exhaust of correction engine.
3. Compressed gas bottles for stabilisation.
4. Radiator for temperature control.
5. Panel of silicon cells which convert sunlight into electricity.
6. Earth directional antennae.
7. Solar panel (see 5).
8. Panel for heating installation. Converts solar rays into heat for apparatus.
9. Solar panel (see 5).
10. Sun directional device for orientation.
11. Base of cylinder, containing camera and other equipment.
12. Section containing gyroscope for stabilisation.
13. Hinge of directional antenna.
14. Reserve antenna (similar to 6).
15. Solar panel (see 5).
16. Inner section of (folding) solar panel (see 5).
17. Small stabiliser rocket, using compressed gas from 3.
18. Pressurised compartment containing radio and other apparatus.
19. Temperature regulating system.

of kilometres of new cable and thus saves an incredible amount of expensive non-ferrous metal and insulating material."

Could Russian engineers who, as early as 1930 were involved with the question of communication, ever have suspected that their problems would one day be solved via space?

COSMIC BIOLOGY

Dogs jump in space – Laika in Sputnik-2 – Experiments with Vostok – Dangers of Cosmic radiation – Two dogs make record flight – Anonymous travellers in space – Day and Night in space

Dogs jump in space

Towards the end of the forties most experts were convinced that from a technical point of view space travel would present no insurmountable problems. The war had produced powerful rockets, capable of conveying heavy loads to a predetermined point many hundreds of kilometres away.

In the final analysis it would merely be a question of time before speeds could be increased to such a point that the payload would end up, not on some point on earth, but in orbit round our planet. In fact it amounted to increasing the velocity from 3 km per second (the velocity of a V-2 engine), to 8 km per second. The solution lay in the multi-stage rocket.

It was an entirely different matter whether living creatures would be able to stand up to the very special conditions presented by spaceflights: the enormous acceleration on launching, the abrupt onset of weightlessness, the dangers of space, such as meteorites, radiation, etc., the absence of the familiar alternation of day and night, the deceleration on return to earth.

These problems were attacked by the Soviets in the late forties, and in many respects they were pioneers in this field.

The Russian biologists intended to launch animals in rockets such as the V-2A and the V-5V, but before doing so they wanted, as far as possible, to carry out extensive experiments in ground laboratories.

Initially research was concentrated on the way animals (dogs, rabbits, rats) reacted to rapid acceleration and deceleration, such as occurs at the launching of a rocket and the re-entry of the nose-cone into the earth's atmosphere. The physiological effect is the same in both cases.

Tsiolkovsky in his time had indicated how it would be possible to imitate the acceleration occurring at the start of a rocket's flight, namely by means of a centrifuge. Dogs were placed in a centrifuge which was spun at a rate sufficient to make the animals feel ten times their own weight. These experiments showed, among other things, that a living creature can stand up to acceleration better if the forces of acceleration operate at right angles to the creature's spine. This is why animals as well as cosmonauts are now launched in a reclining position. Dogs were also subjected to vibrations and noise, such as occur on a rocket's lift-off.

Reaction to decreasing atmospheric pressure was registered in vacuum chambers simulating space. Prototypes of spacesuits were also tested in these chambers. These suits replace atmospheric pressure and thus prevent dangerous effects, such as the formation of air bubbles in the bloodstream. The boiling point of a liquid decreases relative to the pressure exercised on it. At a height of 15 km blood will boil at normal body temperature.

The significance of these experiments became obvious when animals were used for actual flights. The irregularities in pulse and breathing noticed during the experiments in the centrifuge were observed equally during lift-off.

The first experimental flights were carried out in 1951 and greatly advanced the modern science termed "cosmic biology". Naturally these flights were limited in scope. The animals were launched in ballistic trajectories in the uncoupled nose-cones of single-stage rockets, such as the V-2A and the V-5V. When the rocket fell away, the animals became weightless for a few minutes, until they re-entered the atmosphere at increasing speed. One might say they "made a jump in space", an operation which could be compared with that of the first two American astronauts, Shepard and Grissom, several years later.

These pioneer flights took place over the Kapustin Yar base near Stalingrad. Each "animal crew" consisted of two dogs, dressed in spacesuits and housed in special containers which formed part of the detachable nose-cone of the rocket. The nose-cone was not hermetically sealed, so that the spacesuits were no superfluous luxury.

During the propelled phase of their flight the animals experienced up to five times their own weight (5 g). The engine then cut out and weightlessness began. The rocket continued

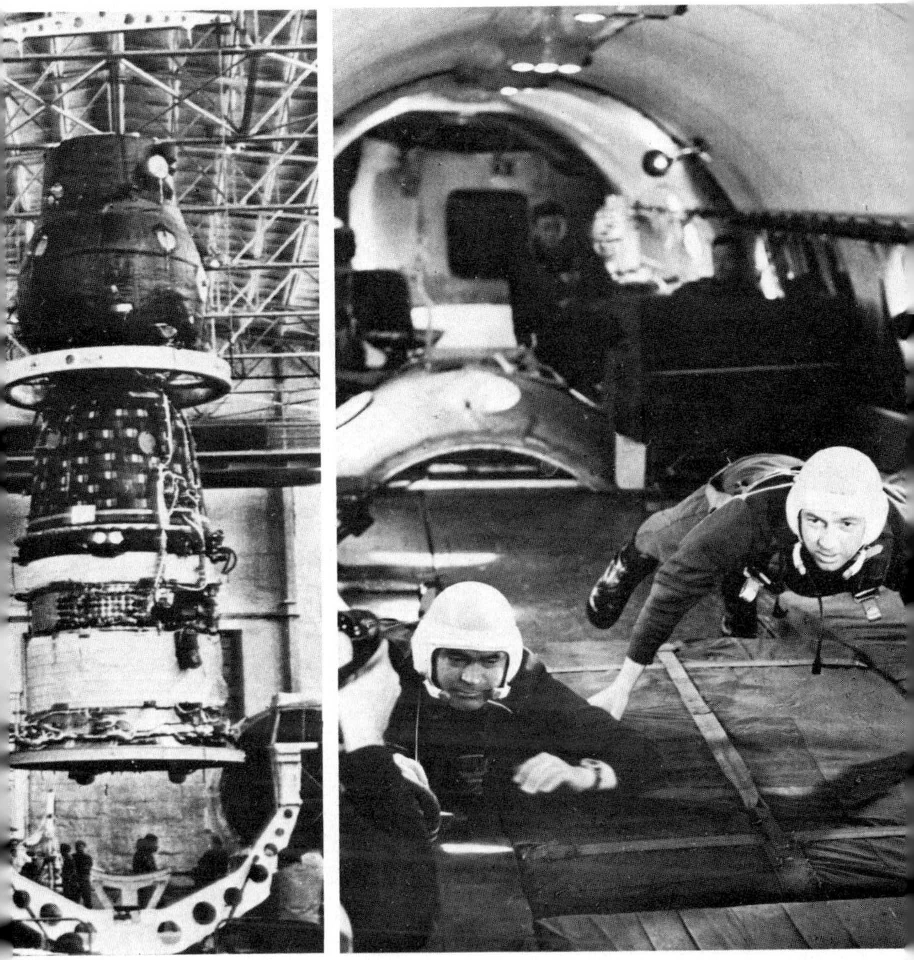

17. *Left* Soyuz-9 in the assembly building at Baikonur. *Right* Andrian Nikolayev and Valeri Sevastyanov experiencing weightlessness in a plane in a parabolic trajectory. Nikolayev and Sevastyanov remained in space in Soyuz-9 for no less than 17 days and 7 hours.

18. *Right* Top of the Soyuz-9 rocket, clearly showing the rockets for detaching the capsule from the carrier rocket in case of an emergency during lift-off. The cylindrical upper section contains small rockets which remove the actual escape rocket as soon as it has become superfluous, that is, when the rocket has reached the upper layers of the atmosphere and the nose-cone separates, releasing the capsule.

(*Below*) Soyuz-9 rocket and spacecraft in transit.

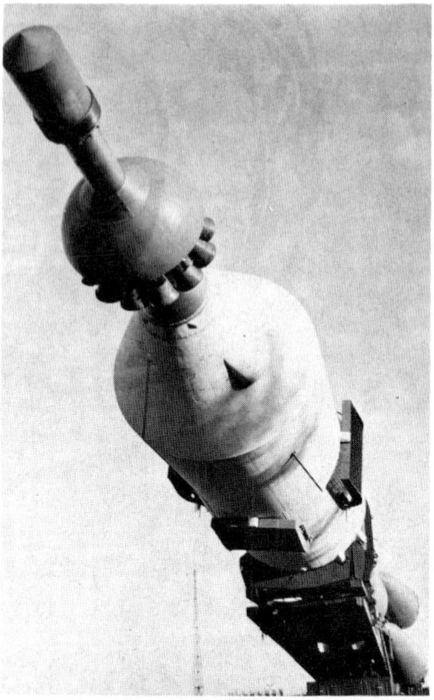

under its own momentum to the apogee of its trajectory, then fell away and back to earth. The nose-cone with the dogs dived separately and at increasing velocity. Forces of deceleration occurred on re-entry into the upper layers of the atmosphere and the animals once more felt the effects of gravity. The maximum period of weightlessness was 220 seconds. In the vacuum of space the dogs' spacesuits provided pressure on their bodies of between 450 and 470 mm (just over half an atmosphere), and the animals were supplied with pure oxygen.

The first dogs reached a height of 70 to 80 km. Later experimental animals reached a height of 450 km, considerably extending the period of weightlessness. On re-entry through the atmosphere, the capsules were decelerated first by means of steel air brakes, then by means of parachutes.

The information thus obtained on the effect of weightlessness —information which could not have been acquired in any other way—was of particular importance. It proved that living creatures could survive zero gravity for a short period at least.

The experiments also provided greater insight into the manner in which the body reacted to psychological influences. For this purpose anaesthetised dogs were sometimes used, who reacted mechanically. Unlike those who were fully conscious during the experiments, these animals showed hardly any irregularity in heartbeat or breathing rhythm.

Automatic film cameras recorded the events on board the capsule. One striking phenomenon was that, immediately upon becoming weightless, the dogs rapidly shook their heads. The cause appeared later: normally the neck muscles are tightened up to a point, in order to carry the weight of the head, and they maintain this tension for some time after the head has become weightless.

Fig. 10. V-2A rocket. (*Drawing and captions overleaf.*)

This rocket has been used since 1949 for geophysical and biological research. The one illustrated overleaf carries two dogs, which, together with the upper part of the rocket, return safely to earth by parachute, after a ballistic flight ("space jump"). The instrument section weighs 1340 kg, the dogs' container 860 kg. The length of the V-2A is 20 metres, diameter at its widest point is 1·66 metres. The rocket was sometimes launched with two instrument containers alongside. When carrying a payload of 2200 kg, it can reach a maximum height of 212 km. This was achieved in 1957.

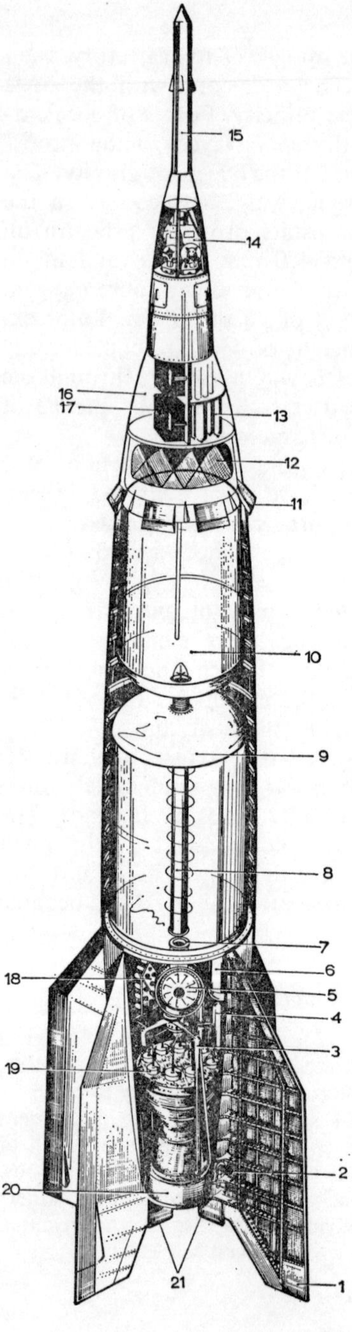

V-2A rocket

1. Aerodynamic rudder of stabilising fin for auxiliary course correction in the atmosphere.
2. Electric motor controlling aero-dynamic rudder.
3. Alcohol supply line for engine cooling. Fuel passes through first section of exhaust and combustion chamber.
4. Permanganate tanks.
5. Turbopump for fuel supply.
6. Hydrogen peroxide tank.
7. Aperture for liquid oxygen supply.
8. Aperture for alcohol supply (twin-walled).
9. Top of oxygen tank.
10. Alcohol tank.
11. Brake flaps for return.
12. Parachute for return.
13. Nitrogen bottles.
14. Cabin holding experimental animals.
15. Tube containing measuring instruments for atmospheric research.
16. Detachable nose-cone (of return section).
17. Radio- and directional apparatus.
18. Air bottles.
19. Injection ducts in engine cover.
20. Engine exhaust.
21. Graphite rudders.

Naturally it was a disadvantage that the dogs had no freedom of movement within their restricted space, since consequently it could not be demonstrated how zero-gravity affected their movements. This aspect had to be cleared up by flights with rats and mice. It turned out that the animals' movements became extremely uncontrolled and unco-ordinated as soon as they were weightless. Obviously they had lost their sense of direction. They subsequently recovered and adapted themselves to the new conditions.

In this connection an important discovery was made by the Soviet biologists E. Yugarov, D. Afanasiyev and J. Kassian. They noted that the disorientation effects ceased as soon as the mice were subjected to artificial gravity by turning their cage. Even mice whose balancing organ had been removed managed very well in weightless conditions.

Soviet experts still maintain—and with justice—that artificial gravity is essential for prolonged spaceflights, for instance in permanent space stations in earth orbit.*

Laika in Sputnik-2

The Russian dogs' space jumps were always of short duration. Interplanetary voyages, on the other hand, would take weeks, if not months. It was therefore only natural that the reactions of an animal in an artificial satellite in earth orbit for hours, or even days on end, would be of great interest. This possibility became reality on November 4, 1957, when the bitch Laika entered Sputnik-2 for what Soviet papers called "a voyage to the stars". Laika was the first living creature to make a true spaceflight.

Although at first the world was left in uncertainty on this point, it was never intended to bring the dog back to earth. At that time the means to do so were not available. The "re-entry problem" had by no means been solved. It was therefore all the more essential to obtain reliable information concerning the pioneer in space by telemetric methods. Laika's cabin in Sputnik-2 was a cylindrical container filled with fresh air which was constantly renewed. The carbon-dioxide and

*See "A Firm Foothold in Space", p. 238.

water vapour exhaled by the dog were sucked out and reconverted into oxygen. Small ventilators provided the necessary air circulation. The temperature in the cabin was maintained at a pleasant level. Laika's food consisted of a kind of paste which contained all essential ingredients, and her excretions were removed via a rubber reservoir attached to her body. The dog's movements were restricted; she was held down by metal bands, mainly to ensure that the many connections attached to her body remained intact. Laika could, however, stretch her legs whenever she wanted to.

Her body was covered with "sensors" which enabled doctors on earth to monitor her continually. Silver electrodes inserted subcutaneously made it possible to make a cardiogram. The breathing sensor was attached by means of a kind of harness round Laika's chest. Even the dog's movements could be "observed" on earth, not via television, as on subsequent spaceflights, but by means of ingenious "feelers" attached to her skin.

During take-off atop the RNV-rocket, Laika behaved according to expectations. She was a little frightened and her pulse and breathing frequency increased. Cardiac reactions remained normal even during maximum acceleration.

Then came weightlessness. Before the flight there had been considerable argument among scientists as to how the animal would withstand a prolonged period of weightlessness. Laika proved to have no difficulty at all. During the ten days she remained alive her reactions continued to be excellent. Then the oxygen supply ran out and she became the first creature to die in space.

Tests with Vostok

It now became desirable to test the reactions of living creatures returning from a spaceflight. The recovery procedure was still regarded as the most difficult phase of the flight and the problem would have to be fully mastered before a cosmonaut could be launched.

There is a considerable difference between the recovery of an unmanned satellite and the safe return of a manned spacecraft. Living creatures are fragile and can survive only limited conditions of temperature and gravity.

The spaceship itself can successfully withstand temperature variations of hundreds of degrees, but a living being can only survive variations of a few dozen degrees. The same applies to the deceleration forces which occur on re-entry. The spacecraft can be made strong enough to cope with these forces, but animal organisms can stand relatively few "g's" within a short period.

Preparation for a spaceflight with a recoverable capsule started on May 15, 1960, with the launching of Sputnik-Spacecraft-1. This was in fact the test flight of an unmanned Vostok, but the world did not know this at the time. It was, however, announced that the new spacecraft weighed no less than 4540 kg, and this in itself was enough to make experts in the West break out into a cold sweat. It was also the first time that a complete Vostok-rocket was used. Sputnik-Spacecraft-1's trajectory ran between 312 and 369 km.

Naturally it was intended to return the new spaceship—which contained a dummy cosmonaut—to earth, but this did not succeed.

At eight minutes past five a.m. on May 19, the retro-rocket was fired, but at that moment the spacecraft was in an incorrect attitude, so that its speed increased rather than decreased. As a result its trajectory was extended to between 307 and no less than 690 km. Part of this spacecraft burnt up after 844 days, on September 5, 1962 when the perigee of its orbit had become increasingly lower as a result of atmospheric resistance.

Fortunately Sputnik-Spaceship-2, which was launched on August 19 of the same year, was considerably more successful. There were living creatures on board: the dogs Belka and Strelka, as well as a number of rats, mice, microbes, seeds and plants. Television was used for the first time to observe the passengers' reactions, but naturally other details, obtained by telemetric methods, remained of equal importance. After 25 hours and just over 17 orbits the living cargo was recovered in good health.

Chief Designer Korolyov and his team, as well as the biologists, had every reason for satisfaction. This was their first opportunity to record the biological effects of every phase of the flight, and—what is of the greatest importance—the reactions of the animals *after* their flight could be studied.

The crew of the "ark in space" had been carefully selected.

Organisms at widely differing levels of evolutionary development—from very simple to very complex—had purposely been chosen. It had already been established in the laboratory how these forms of life reacted to radiation such as occurs in space. Black mice were used because their colour changes rapidly when subjected to strong electromagnetic radiation. It was an encouraging fact that the rats and mice had eaten well in spite of floating weightless in a very large cage. This indicated that they had adapted to weightlessness. During the short ballistic flights of the early fifties the animals had always been panic-stricken and had left their food untouched.

After the flight all passengers proved to be in excellent condition. Both Belka and Strelka subsequently produced several litters of perfectly healthy puppies.

The fact that one swallow does not make a summer was proved once again with the next flight: Sputnik-Spaceship-3. The passengers in this capsule, including the dogs Ptsyolka and Mushka, did not return alive. The spacecraft and its crew completed seventeen orbits in perfect condition, but the angle of its re-entry was so steep that the capsule burnt up.

Sputnik-Spacecraft-4, launched on March 9, 1961, was considered by the Russians as the direct precursor of the first manned spaceflight. The capsule containing the dog Chevnushka completed exactly one orbit at between 183·5 and 248·8 km and then landed safely "within the predetermined area". The objective of this test flight was not so much to obtain extensive biological data, but rather to bring the dog back in good condition from an elliptical orbit. The perigee of the orbit had purposely been fixed so low that atmospheric resistance would return the capsule to earth within ten days. The passenger would still be alive, since food supplies had been calculated for this period. This was a precautionary measure, in case the retro-rockets should not function. A month later Gagarin was to complete a similar orbit. "It is five minutes to twelve", the West concluded when Chevnushka's safe landing became known.

Man's final pathfinder was Zvezdochka (Little Star), who was given this name shortly before setting off in Sputnik-Spacecraft-5 by the man who was to follow in her footsteps: Yuri Gagarin. After one orbit, Zvezdochka, too, landed in perfect health. That was on March 25, 1961. Now it was one minute to twelve. . . .

Dogs are tops

It was no mere coincidence that Russian scientists chose dogs for experiments in space, while the Americans concentrated on monkeys in their preparation for manned spaceflights. Since Ivan Pavlov at the turn of the century for the first time made an in-depth study of the reactions of a dog, the Russians have acquired greater knowledge about this animal's physiology than anyone else. Dogs moreover are exceptionally suitable for experiments in space. Their heart and blood vessels, for instance, are sufficiently similar to man's to make comparison valid. Moreover dogs are undeniably intelligent and suited for specific training, and they possess the inestimable ability to adapt easily to changing conditions. The Russians are therefore correct when they say that dogs make ideal pathfinders for man.

Once a human being had set foot in space—albeit within the safe confines of his spacecraft—it was tacitly expected in the West that the Russians would no longer use dogs as travellers in space. This was far from being the case. A Soviet expert once said: "Dogs are the explorers of space. They will always be the first to explore new, uncharted routes. But animals are not only explorers. Some of them have been man's friend for thousands of years and they will certainly be of use on spaceflights. Some animals and plants will form part of the miniature biological cycle on board a spaceship, and on prolonged spaceflights man will use them as food."*

The use of animals as biological "alarm signals" has often been suggested in professional Russian journals. Animals can sometimes detect certain dangerous conditions better than man. A well-known example is the fact that miners used to take canaries underground, since these were exceptionally susceptible to fire-damp.

Russian scientists have more than once mentioned the possibility of using algae for producing oxygen in the spacecraft and space stations of the future. The problem of how to carry sufficient oxygen for the crew, even in liquid form, becomes increasingly difficult on more prolonged spaceflights. It will become essential continuously to release oxygen from the carbon-dioxide and water exhaled by the crew—something

*See "A Firm Foothold in Space", p. 238.

which has so far not been done. Algae consume carbon-dioxide and can rapidly release large quantities of oxygen. Therefore the Russians, who always show a tendency for using "household remedies", consider algae very suitable. The Americans, on the other hand, though not denying the theoretical importance of algae, nevertheless prefer chemical apparatus to refresh the same air again and again.

Dangers of radiation

Where space travel is concerned, the Russians have some specific preoccupations. Soviet space biologists have more than once given the impression of being pessimistic regarding the problem of cosmic radiation. Part of this radiation emanates from the Milky Way, part from the sun. The latter source is the more variable. At times fierce explosions occur on our central star, when hydrogen and helium nuclei, as well as nuclei of heavier atoms are emitted into space. It is particularly the heavier nuclei which are the culprits: although relatively few in number, it is they which chiefly (for 80 per cent) determine the biological effect of cosmic radiation. Space travellers subjected to excessive radiation may suffer genetic injuries which may adversely affect their progeny.

Fortunately it is now possible to predict coming solar explosions with a fair degree of accuracy at least a few days in advance. This is sufficient for fairly short voyages in space. The difficulty would arise with a voyage to Mars (8 months!) or with a prolonged stay in a space station orbiting the earth at high altitude.

The Soviet cosmonauts never went higher than 500 km on their flights round the earth. The medical men connected with the space programme pointed out the reason for this: the Van Allen radiation belts. These consist of high-velocity nuclear particles, originating in cosmic radiation, caught in the earth's magnetic field. The density of these belts varies with the activities of the sun.*

The dose of radiation to which a cosmonaut may be subjected during flight in these belts varies between 75 and 100 röntgen per second, a fatal dose, for not more than 15 to 20

*See also: *Elektron*, p. 86.

röntgen per day is admissible. This is why the Russians are apprehensive of flights in much higher orbits. The problem is less severe where flights to the moon are concerned, when the belts of radiation are traversed vertically at great speed, preferably at a time when the sun is relatively inactive.

The Russians regard cosmic radiation—both within the belts and in free space—as a serious problem. Viktor Bazykin, a member of the Association for Astronomy and Geodesy of the Academy of Sciences, once wrote: "The information obtained with the aid of Sputniks, rockets and balloon probes has proved beyond all doubt that radiation is the chief obstacle to penetration into cosmic space".

Needless to say during manned spaceflights the Russians constantly and carefully monitor radiation conditions. They do this not only from earth, but also by means of Kosmos satellites. Increasing frequency of Kosmos launchings can therefore often be interpreted as an indication that a new manned space project is being planned.

Meanwhile a great deal of work has been done on devising means to protect man from the effects of radiation. On the one hand it is possible in principle to shield spacecraft from penetrating radiation. On the other hand, drugs have been developed in the Soviet Union, which lessen the effects of radiation.

In theory the first method is of course preferable, at least as long as the cosmonauts remain inside their capsule. In practice, however, there is one great disadvantage: an effective shield is very heavy, so heavy in fact, that economies have to be made in the weight of the remaining equipment if the craft is to be got off the ground.

It is therefore logical that the Russians have been greatly preoccupied with the second method: drugs. Cosmonauts carried these drugs on all flights, in case a sudden increase in solar activity should make it desirable to use them, after which they would have to return to earth forthwith. In practice, however, such an emergency has not yet arisen.

Record Flight

The Russians would be the first to admit that medical science has by no means solved the problem of dangerous radiation.

It is therefore not surprising that, in preparation for the flights which would take the cosmonauts further into space, a special satellite was launched to find out what could be done to counteract intensive radiation. This was Kosmos-110, in which the dogs Veterok and Ugolek circled the earth at a height varying between 187 and 904 km for no less than three weeks. Up to that time this was the longest space flight ever made by a living creature.* In the upper levels of its trajectory, Kosmos-110 obviously traversed the offshoots of the inner radiation belt, and it was therefore interesting to observe the effect on living creatures existing in these regions for some weeks. Apart from the dogs, the satellite contained spores of yeast mould, blood serum, albumen preparations, chloral-hydrates and bacteria.

The investigations were not confined to radiation alone. The effects of prolonged weightlessness on heart and blood circulation were also studied. Veterok (Breeze) was the chief guinea-pig. By means of automatic apparatus the dog was regularly given drugs against the effects of radiation, whereas the dog Ugolek, for purposes of comparison, received no such medication. Continuous cardiograms and blood pressure readings were taken of both dogs.

As for radiation, research also took place into the possibility of allowing the wall of the spacecraft to absorb as much radiation as possible without resorting to an actual "shield". Of course the other living material as well as the dogs was studied in relation to radiation effects.

Immediately after the landing of Kosmos-110, on March 17, 1966, Yegorov (the doctor who had previously been a member of the Voskhod-1 crew) and his colleagues carried out a preliminary examination of the dogs. "Our immediate observations led us to the conclusion that a living organism can survive a prolonged period of weightlessness", said Yegorov. About the effects of radiation he said: "Kosmos-110 definitely passed through the radiation belt. The information transmitted to earth requires further study. In any case we may say that at first sight we have noted no harmful effects. The electrocardiogram which was made of the dogs today looks good." It was several weeks before the first results of the experiment were published. Evidently the dogs (nothing was said about the rest of the living cargo) had certainly been affected by their stay in

*Their achievement was surpassed only by the crew of Soyuz-11-Salyut.

space. It appeared that immediately after their 22-day space marathon they had presented dehydration symptoms.

Similar symptoms had previously been reported by American astronauts. They had been instructed to drink as much as possible, even when they did not want to. The fact that Ugolek and Veterok were somewhat weak after such prolonged weightlessness is hardly surprising.* People who have been on their backs for a few weeks may also be dizzy and weak-kneed for a while. More serious was the phenomenon—already noted on previous spaceflights—that during the period of weightlessness blood and urine absorb calcium from the bones. The American astronauts tried to prevent this by doing physical jerks to exercise their bones. The dogs, of course, did not exercise, so it was not surprising that in their case decalcification was very pronounced. It is well known that a loss of calcium occurs to a smaller degree as a result of a prolonged stay in bed, when there is little bone function. As a matter of fact, the condition of Nikolayev and Sevastyanov, who orbited in Soyuz-9 for eighteen days, could hardly be called ideal, in spite of the fact that they had exercised intensively.

Oddly enough the first brief Russian reports of the Kosmos-110 experiment did not mention the most interesting aspect of the flight, namely that the living cargo of this Sputnik had flown for more than three weeks through the offshoots of the notorious radiation belts. It was, however, stated specifically that both dogs were in excellent health. It is difficult to say whether the experiment convinced the Russians that radiation presents no actual danger for flights deeper into space, particularly to the moon. It is, however, certain that the Americans were not hindered by radiation on their Apollo flights.

Anonymous Space travellers

People as well as dogs acted as guinea-pigs in the Russian preparations for manned spaceflights. Dozens of men and women in the USSR volunteered for extensive experiments before man set foot in space. Most of them remained anonymous and none of them ever flew in space. They were

*See "Marathon in Space", p. 186.

spun in centrifuges until they lost consciousness, they spent hours in vacuum chambers to test prototypes of spacesuits, their powers of endurance were tested to the utmost in intense heat. In this way the medical men gained insight into the limits of tolerance of the human organism, limits which could not be exceeded, particularly in those phases of the flights where the system is subjected to the greatest pressures, namely the launching and the re-entry into the earth's atmosphere.

Day and Night in space

Another important factor affects the reactions of living organisms during flight in space. Up to now little attention has been paid to this point, but it is gradually being realised that it is something which may have a distinct effect, particularly on prolonged spaceflights: the absence of the day/night pattern and the problems this brings with it.

The Soviets were the first to make a point of this. In each human being a mechanism is concealed which so far has been little understood, and which is linked to certain natural rhythms: day and night, the month, the seasons. The day-and-night mechanism is the most obvious, since it is the most frequent. It is sometimes said that a "biological clock" governs our bodily processes. This explains why one does not feel well when the rhythm is broken in some way, for instance when one sleeps during the day and works at night. Similar symptoms occur, often in very unpleasant forms, in people who fly in jet planes from east to west or vice versa. Because they move so rapidly, they fly towards the sun, or remain ahead of the sun, so that their day is suddenly considerably shortened or lengthened. In the long run this may lead to the "jet-disease" which in turn may result in all kinds of organic disturbances.

A cosmonaut who circles the earth in a relatively low orbit sees the sun rise and set every 90 minutes. This does not make him feel that the day lasts only an hour and a half, for, whether the sun is in the sky or not, in the absence of the atmosphere which refracts the blue element of sunlight most strongly, the sky is black.

One might therefore expect that it is not very important how a cosmonaut divides his time, provided he has sufficient rest. Experience, particularly on the first prolonged American spaceflights, has shown that this is not so. It is of the greatest importance that the 24-hour cycle should be adhered to as much as possible. Whenever practical, the cosmonaut should work and sleep at the same times to which he is accustomed on earth. The Russian professor Parin,* among others, has said more than once that this question is regarded as very important. The Americans have reason to agree, for during their first few Apollo flights they experienced some problems in this respect, as they had not taken the terrestrial cycle into consideration and moreover gave their astronauts too little rest. This changed with Apollo-10 and since then there have been no problems on this point. Obviously even the most rigorous training cannot overcome the experience pattern produced by millions of years of human evolution.

Not only is there a day and night cycle, there is also a monthly one. The moon, which circles the earth every four weeks, has a greater influence on life than we perhaps suspect. It causes tides not only in the seas of our planet, but also in its atmosphere and consequently influences numerous natural events. After millions of years of evolution this influence has become quite obvious, for instance in the monthly periods of women. This was one factor which had to be taken into account in the preparations for Valentina's flight. On this subject space doctor Yazdovsky said:

"The female cosmonaut had to undergo the same training as her male colleagues. However, it was necessary to take into account a woman's anatomical and physical characteristics, particularly the changes in reaction at various times of the menstrual cycle.

"Special experiments were carried out with female rhesus monkeys. Research into the effect of rapid acceleration showed that the changes were least during the fertile period of the menstruation cycle, with very slight changes during ovulation. This information was used in the development of suitable centrifuge training and the determination of the optimal times for a woman's flight in space."

When Valentina Tereshkova was launched on June 16, 1963, few people could have realised that the date was closely linked with the physical condition of this first female space traveller. In the Soviet Union, therefore, as little as possible is left to chance, particularly where human spaceflight is concerned!

*Professor Parin died in June 1971.

VOSTOK PAVES THE WAY

Gagarin: first man in space – Six hundred thousand kilometres through the universe – The first "cosmic twins" – Valeri and Valentina: man and woman in space.

Around the world in ninety minutes

At Baikonur the weather is poor for at most five days in any year. It is therefore no coincidence that at dawn on April 12, 1961, warm sunshine greets the scientists, technicians and workmen who have been working all night on the assembly platform surrounding the slender white rocket bearing the Cyrillic letters CCCP in red.

At the same time, in Zvyezdograd (Star City), a man is woken from a deep sleep by his "personal physician":

"Yuri, it's time to get up."

Within a second Yuri Gagarin is wide awake. "Get up? Sure!" Almost simultaneously he and his stand-in, Herman Stepanovich Titov, jump out of bed.

"How did you sleep?" asks Dr. Yevgeni Anatolyevich.

"As I was taught to," Gagarin says, smiling.

After their morning exercises and a wash they breakfast from tubes: Meat paste, marmalade, coffee. After this follow medical examinations and a check of the instruments which later on, from space, will transmit information about heart beat, breathing, and blood pressure. With the help of their colleagues, who have meanwhile arrived to be present at the historic launching, Yuri and Herman get into their space-suits: first a pale blue pressurised suit, then a scarlet overall. Finally the two space travellers put on the cap with earphones, and the white helmet bearing the letters CCCP. They are ready for the start! In a special coach the two pioneers are driven to the launching pad, where meanwhile technicians have completed their final intensive checks. The assembly towers have been pulled away. Only now the slender shape of the rocket with its four large boosters can be clearly seen.

Dozens of well- and lesser-known personalities are assembled on the base, the chief designer, Dr. Yevgeni and some members of the State Commission for the Achievement of the First Manned Spaceflight.

It is seven o'clock Moscow time when Yuri Gagarin stands on the platform, facing the entrance to the spaceship. With some difficulty—the spacesuit makes him clumsy—he lifts both hands in farewell to his stand-in and the others who will stay behind. Then the cosmonaut disappears into the capsule.

Hardly has Vostok,* as the spaceship is called, been hermetically sealed, than Gagarin's voice becomes audible in the control centre:

"Attention earth, here is the cosmonaut. Radio connection tested. Initial position of switch on guidance system panel correct. Globe in starting position. Cabin pressure 1, humidity 65, temperature 19 degrees, cell pressure 1·2, pressure in orientation systems normal. Am feeling fine. Ready for lift-off."

A few more minutes. Now the commands from the firing officer, who is watching the rocket through a periscope, follow rapidly one after another:

"Klutsh na start!"	(Switch in starting position. The automatic count-down apparatus is now functioning).
"Protyazhka odyin"	(First automatic systems check).
"Produvka"	(Blow-through. Combustion chambers and pipelines are flushed with nitrogen).
"Ventilyatsiya"	(Ventilation of fuel tanks to prevent excessive pressure).
"Klutsh na drenazh"	(All valves closed).
"Pusk"	(Ready for lift-off).
"Protyazhka dva"	(Second automatic systems check—the gyroscopic steering system of the rocket is now functioning).
"Zemlya-bort"	(Connection between cablemast and rocket broken—the fuel valves are open, the turbines are working).
"Start"	

Will Gagarin be able to withstand the weightlessness to which he will be subjected as soon as his Vostok has gone into orbit? That is the question which most occupies the scientists below.

"At first it was a strange feeling," Gagarin said later, "but

*"Vostok" means "East".

Fig. 11. Vostok spaceship. (External diameter of the capsule: 2,300 mm.).

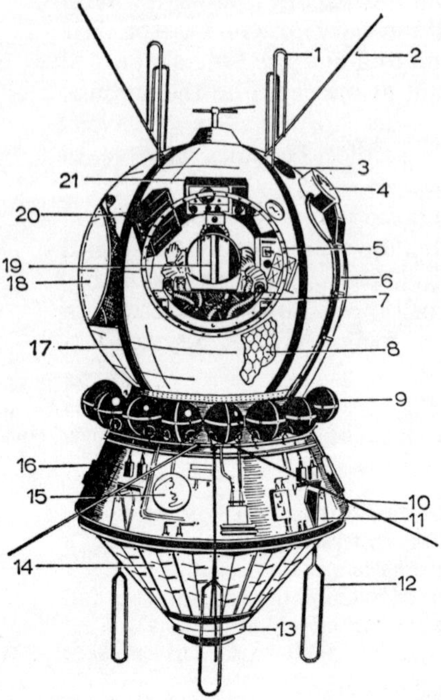

1. Radio antennae.
2. Communications antennae.
3. Capsule porthole.
4. Multi-adaptor plug for connections between capsule and instrument section.
5. Opening for hatch-cover. Explosive bolts jettison hatch cover during parachute descent of capsule, so that ejector seat can come out.
6. Recessed cable.
7. Ejector seat with parachute and cosmonaut.
8. Honeycomb heat shield (exposed).
9. Gas containers for steering rockets etc.
10. Antenna for communication with earth.
11. Antenna for orbit measurements.
12. Antenna (telemetry).
13. Retro-rocket.
14. Shutters for temperature control.
15. Hatch to instrument section.
16. Instrument section.
17. Steel bands connecting capsule to instrument section.
18. Parachute hatch.
19. Cosmonaut's helmet.
20. Switch panel.
21. Instrument panel.

I soon got used to it and could work normally. Everything was suddenly easier. I felt as if my arms and legs, in fact my entire body, no longer belonged to me. I was not sitting or lying down—I seemed to hang somewhere in the air. All unsecured objects were floating around and when I looked at them I seemed to be dreaming. There flew the atlas, the pencil, the notebook."

Gagarin expressed the opinion that gravity does not affect a man's capacity to work. Scientists preparing for longer voyages in space need not worry.

Gagarin's flight lasted only 108 minutes; he completed

). Soyuz-9 being prepared for lift-off. The rocket is surrounded by retractable
ıounting platforms. In the foreground is the transporter which has conveyed the
ɔcket to the launching pad.

20. The nocturnal lift-off of Soyuz-9.

one orbit round the earth. Vostok's retro-rockets were fired over Africa at 10.25 and the spherical capsule of the spaceship landed with the aid of an enormous scarlet-edged parachute in the neighbourhood of Saratov, on the banks of the Volga. It was exactly 10.55 when Gagarin came down.

Tass announced:

"*After the successful completion of the planned investigations and of the flight programme, the Soviet spaceship Vostok made a safe landing at the predetermined place on April 12, 1961 at 10.55 Moscow time.*

The pilot-cosmonaut Major Gagarin made the following statement: 'I beg to report to the Party, to the Government and to Nikita Sergeyevich Khrushchev personally, that the landing was normal, that I am feeling well and have sustained no injuries or disturbances.'

The accomplishment of a manned spaceflight opens up grandiose perspectives for man's conquest of the cosmos."

Shortly afterwards a biography of the brave space traveller was published, which showed that Gagarin was not just a "simple metalworker", as some western papers would have their readers believe.

Yuri Gagarin was born in a small town near Gzatsk on March 9, 1934, the second son of humble parents. Until June 22, 1941 he attended primary school. On that day Hitler invaded the Soviet Union and all schools were closed; they did not re-open until 1945. At the same time Yuri joined the Pioneer Organisation, which may be compared with our Boy Scout movement. During this period he read a book by the great pioneer of space travel, Konstantin Tsiolkovsky, and he became enthusiastic about the conception of space travel. After the six-year course at the Gzatsk school he joined the Technical College. That was in September 1949. In 1951 he obtained his school-leaving certificate—with credit.

Yuri then moved to Saratov (in the neighbourhood of which he was to land after his spaceflight) to attend the Industrial Technical College, where he specialised in metalwork. In this period he joined the Aero-Club, an association of young amateur flyers. In June 1955, while space experts and the military authorities in the USSR were working hard on the first intercontinental rocket, Gagarin, bursting with enthusiasm, made his first solo flight. That same year he joined the Pilot Training College at Orenburg, where he completed his training as a fighter pilot in 1957. In May of that year the young lieutenant made his first flight in a jet fighter. In November 1957, in the exciting days after the surprise launching of the first Sputnik, another important event took place for Yuri. He married Valentina Ivanova, who bore him two children, Lanotshka and Valya. Towards the end of 1959 Yuri heard of the possibility of being trained for space travel. He seized his chance with both hands. His subsequent intensive training finally, on April 12, 1961, made him the first man in space.

Fig. 12. Instrument panel in Vostok.

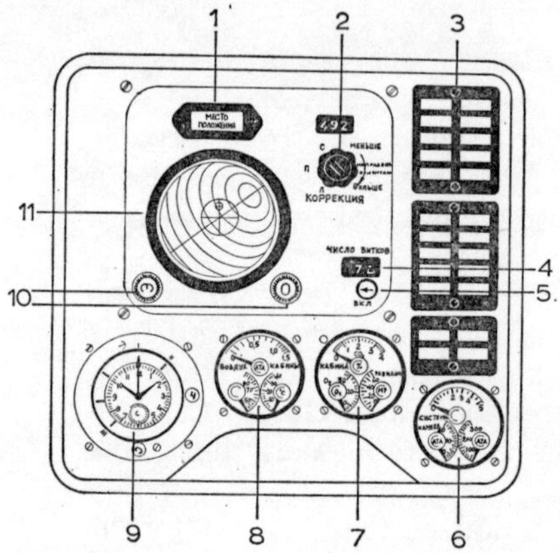

1. Indicator light "position spacecraft".
2. Push button for correcting position of globe.
3. Indicator lights of various systems on board.
4. Number of orbits completed.
5. Push button for orbit counter.
6. Pressure meters of gas containers.
7. Meter for checking carbon dioxide and oxygen content as well as radiation level in cabin.
8. Meter for checking pressure, humidity and temperature in cabin.
9. Chronometers.
10. Globe adjuster knobs.
11. Globe which revolves synchronically with the spacecraft's movement round the earth, so that it constantly shows the cosmonaut his exact position.

At a well-attended press conference, held shortly after the flight, it was announced that the Vostok flight had been controlled by the automatic guidance system, although Gagarin could at any moment have taken over manual control. He could also have carried out the landing procedure. Although there are conflicting reports on this point, it seems probable that Gagarin landed in the capsule, although he could have been ejected in his ejector seat at a height of some kilometres to land, like the spacecraft, by parachute.

Some of the questions posed at the press conference were

very interesting indeed. The same cannot be said, however, of all the answers, but in many cases the fact that they were non-committal helped to improve the atmosphere in the hall of the crowded Palace of Science.

The following are some of the questions and answers on that occasion. (The questions had to be written down and handed in beforehand; many remained unanswered.)

Question: When were you told that you were the first candidate?

Gagarin: I was told in good time that I was the first candidate.

Question: You said yesterday that your fellow-pilot-cosmonauts are prepared for another cosmic flight. How many are there? Are there more than a dozen?

Gagarin: In accordance with the plan to conquer cosmic space, pilot-cosmonauts are being trained in this country. I believe that there are more than enough to undertake important flights.

Question: What is your salary? Have you received any special reward for your flight?

Gagarin: As is the case for all Soviet citizens, my salary is more than sufficient to meet all my needs. I have received the decoration of Hero of the Soviet Union. That is the highest distinction.

Question: When will the next spaceflight take place?

Gagarin: I think that our scientists and cosmonauts will undertake the next flight when it is necessary.

However, some of the answers *did* provide some information:

Question: How did you feel when your sense of gravity disappeared on your way up, and when it reappeared?

Gagarin: When I became weightless I felt perfectly well. Everything became easier to do. This is understandable. My hands and my feet had no weight. Objects floated in the cabin. I myself was not sitting in my chair as before—I hung in the air. In this condition of weightlessness I ate and drank as usual—everything was the same as on earth. I worked as well; I wrote and noted down my observations. My handwriting was the same as always, even though my hand was weightless. I did, however, have to hold on to the notebook, as otherwise it would have floated away. I maintained radio communications on several channels. I operated the Morsekey. As far as I could see weightlessness did not affect my capacity to work. The transition from weightlessness to gravitation, to the reappearance of gravity, is very gradual. Hands and feet felt exactly as before I became weightless. I no longer floated above my seat, but sat in it once more.

Question: What do the sunlit and the dark sides of the earth look like? How do the sky, the sun, the moon and the stars look?

Gagarin: The sunlit side of the earth was perfectly visible: the coasts of the various continents, islands, large rivers, large lakes, folds in the earth are all easily discernible. When I flew over our country I could clearly see the large square kolkhos fields; I could even distinguish between arable land and pasture. I had never before been able to get above 15 000 metres. Of course from a sputnik one cannot see as well as from a plane, but it is still very, very good. For the first time I saw the spherical shape of the earth—one can see it by looking towards the horizon. I must say that the horizon presented a very unusual sight. There is a surprisingly colourful transition from the lit surface of the earth to the black sky filled with stars. This transition appears as a thin, bluish belt surrounding the earth. The transition from blue to black is very gradual and is a truly magnificent sight. It is difficult to describe. And when I flew out of the earth's shadow, the horizon looked quite different. There was a bright orange band, shading towards blue and then to black. I did not see the moon. In space the sun is many times brighter than on earth. The stars are clearly visible: they are very bright and distinct. There is much more variation in the dome of the sky than we can see from earth.

Question: It is said in the foreign press that the United States intend to send a man into space as well. What is your opinion on this?

Gagarin: Our party and our government are in favour of peaceful utilisation of space and of peaceful competition. Of course we would rejoice in the successes of American astronauts. There is space for all in the universe. But it must be used for peaceful, never for military purposes. The American cosmonauts will have to overtake us. We shall applaud their successes, but will do our best to remain ahead.

Question: Would you, within the framework of the present programme, be able to fly to the moon in the Vostok?

Gagarin: The Vostok was never intended for a flight to the moon. Special spacecraft will be built for this purpose.

Six hundred thousand kilometres through the Universe

While Gagarin continued his goodwill tour which had begun in space, the Baikonur technicians started preparing for a new flight, one which was to last much longer.

Gagarin's 108-minute voyage had proved that man is

capable of withstanding the forces of acceleration and deceleration during launching and landing, as well as a short period of weightlessness, but it was by no means certain that a human being could remain weightless for hours on end without harmful consequences. It was, moreover, necessary to confirm the subjective opinion of cosmonaut No. 1, that zero gravity does not affect the capacity to work. And what about Vostok's manual steering gear, which had never yet been tested in flight? A simulator is after all not the same thing.

On August 16, 1961, at 5 a.m., Gagarin, who was then on tour in Canada, was awakened by his host, Cyrus Eaton, who knocked on his door and called out excitedly: "Vostok-2 has been launched!"

Gagarin, who at a press conference the day before had feigned ignorance when asked who would be the next man to penetrate the universe, proved to be well-informed. "I know. It's Herman Titov. When we get back to Moscow he will have just returned to earth."

Herman Stepanovich Titov had started at 6 a.m. Moscow time, watched by dozens of technicians as well as by cosmonaut No. 3, Andrian Nikolayev, who was his stand-in. Only a few minutes later Vostok-2 went into orbit at a height varying between 178 and 257 km.

In the course of this flight, which lasted more than 24 hours, Titov saw practically the whole of our planet's surface, for the earth revolved in a west-east direction under the trajectory of his spaceship. Each orbit—that is, approximately every 90 minutes—Vostok-2 overflew a point about 22 degrees further west than the one Titov had seen before him during the previous orbit.

At 10.18 Moscow time on August 7, Titov landed near Saratov, not far from the point where Yuri Gagarin had come down. When on landing he reported to the doctors Vevgeni Anatolyevich Pobydonostsyev and Vladimir Ivanovich Yazdovsky, it turned out that he had felt unwell for a time. He had become dizzy and had had to lie down for several minutes, until his vertigo became less severe. This was the first occurrence of the "space sickness" predicted by the Dutch doctor M. P. Lansberg, who had already described this malady in his thesis "Problems of Space Medicine", with which he obtained his doctorate on June 10, 1958(!).

Titov related that, soon after arriving in orbit, he had for a short time had a sensation as if he were flying upside down. The sensation rapidly disappeared and he could carry out his tasks normally. Later on he became dizzy if he moved his head in certain ways. There were clear indications of problems of adaptation as described by Dr. Lansberg.

It is particularly the inner ear which is involved in this problem of adapting to weightlessness. An important rôle is played by the otoliths, minute bone concretions which press against the wall of the inner ear as a result of gravity and pass information to the brain via the nerves. It is the otoliths which enable us to determine our posture, even with our eyes closed. It is therefore understandable that in conditions of zero gravity, and particularly when the head is turned quickly, a wrong impression may be formed, which may in fact conflict with what the eyes see.

It was not disclosed until several months later that in Titov's case the unpleasant sensations started about two hours after lift-off and reached their maximum effect approximately six hours later. His appetite suffered also. After he had slept, however, he felt much better, particularly since he had learnt to avoid sudden head movements. Similar unpleasant sensations were experienced by other Russian cosmonauts. Less well-trained space travellers, such as the engineer Feokistov and the doctor, Yegorov (Voskhod-1), suffered particularly from space sickness.

At the press conference after his flight, Titov disclosed for the first time that he had landed by parachute. He had ejected himself from the Vostok at a height of seven kilometres. The spaceship itself also came down by parachute, not far from the place where the cosmonaut landed.

The flight of cosmonaut No. 2 had proved that a human being can remain in space for a full day without encountering serious problems. Certainly it had become obvious that the training programme would have to pay more attention to a man's power of orientation, which would lessen the chance of space sickness.

It was Herman Titov who said after his successful flight: "My successor, who is an unmarried daredevil and a great chap, will make a much more sensational voyage in space in the near future. . . . "

Of course he did not add that this voyage would involve a first attempt at solving the important docking problem.

The "Cosmic Twins": Nikolayev and Popovich

Titov's prediction came true on the morning of August 11, 1962. One day later it appeared that the Soviets were making a first attempt at solving the problem of a rendezvous in space.

On August 11, at a time when most Moscovites were having their midday meal, Tass announced:

"On August 11, 1962, at 11.30 Moscow time, the spaceship Vostok-3, commanded by the Soviet Cosmonaut Comrade Major Andrian Grigorevich Nikolayev was put into orbit round the earth. . . . "

Vostok-3 travelled in an orbit of between 183 and 251 km, at an angle of 65 degrees with the equator.

To everybody's amazement a fourth Vostok, carrying 31-year-old Pavel Popovich, was launched less than a day after the departure of the 33-year-old bachelor Nikolayev. In the Tass communiqué released on August 12, the rendezvous problem was referred to in very cautious terms:

"Today, at 11.02 Moscow time, the spaceship Vostok-4, with Cosmonaut Pavel Popovich on board, was put into orbit.

"According to plan, the Vostok-4 launching took place while Vostok-3, launched in the USSR on August 11, is still in orbit.

"At this moment there are two Soviet capsules in cosmic space: Vostok-3 and Vostok-4, manned by the Soviet citizens cosmonauts Andrian Grigorevich Nikolayev and Pavel Romanovich Popovich.

"The purpose of launching two spacecraft in orbits at little distance from each other, is to obtain information on the possibility of *establishing a direct link** between the two capsules; and in addition to let the cosmonauts carry out co-ordinated tasks and compare the effects under similar conditions on the human organism. . . . "

Meanwhile the two cosmonauts continued their triumphant spaceflight. At one moment the two spacecraft were so close to each other that the cosmonauts could see each other's capsule through the porthole; the distance was 6·5 km.

If one considers that this was due solely to accurate launching (the capsule could not be steered), one cannot but admire the achievement.

*Author's italics.

Nikolayev had been launched at 11.30 Moscow time on August 11, Popovich nearly 24 hours later, at 11.02 on August 12. At the moment when Vostok-4 took off into space, Vostok-3 was approaching the Baikonur base.

The capsules would have been even closer together, if Vostok-4's speed had been slightly less. Moreover the angles of their orbits with the equator differed slightly in each case: Vostok-3's angle was 64·98 degrees and Vostok-4's 64·95 degrees.

The rendezvous-at-a-distance took place shortly after Popovich's take-off. The two capsules then moved apart once more, although the two cosmonauts remained in radio contact for quite a time.

During this "group flight" Nikolayev and Popovich, as far as possible, led the same kind of existence, so that interesting comparisons could be made after the flight. Now and then the two men took off their clumsy spacesuits. They photographed the moon and other celestial bodies, they ate, drank, slept and performed tricks for the benefit of television viewers on earth. Before they went to sleep they sang the cosmonauts' song together:

I believe, my friends, that caravans of rockets
Will take us from star to star.
Our footsteps will leave traces
On the dusty paths of distant planets.

On Wednesday, August 15, a Japanese radio station announced that the landing instructions to the two cosmonauts had been picked up. That morning, a few orbits before the time of landing, Baikonur instructed Popovich:

"Mission control here. As your instruments indicate a low temperature and degree of humidity, and in view of the fact that you have carried out your task, prepare for landing in the 49th orbit. Check your safety equipment, the switch of your ejector seat and the condition of your spacesuit. Wind speed at the place of landing is seven to nine metres per second."

Later the calm voice of Nikolayev was heard: "Preparing for landing in the 65th orbit. Cabin pressure is 1·1, temperature 11 degrees Celsius, humidity 70 per cent."

Vostok-4 returned from space six minutes after Vostok-3. Both capsules landed east of the launching base, about 300 km apart.

On the Saturday following the landing, the cosmonauts were given a jubilant reception in Moscow, where they met their relations for the first time after long separation. Shortly afterwards, in a talkative mood, Nikolayev told a reporter: "I'd like to get married. I have no special girl in mind yet, but I should now like to come to a speedy decision. . . ."

The hawk and the seagull

"Hello, earth, here is seagull—hello, earth, here is seagull."

These words, spoken in a high, somewhat excited voice, come over the loudspeakers in the block-house of Baikonur on June 16, 1963.

Shortly before, at 12.30 Moscow time, the Russians have for the first time put a woman into space: Valentina Vladimirnova Tereshkova. Soon after take-off she meets—at a distance of 5 km—Valeri Fyodorovich Bykovsky, who has by then already been in the cosmos for two days and is establishing a record.

"When I heard Valeri's voice, it was just as if we were in the same room," Valentina told a press conference afterwards. It had originally been intended that Tsyaika (seagull) would return to earth after one day, but the flight was going so well that it was decided to extend it to three days. Thus it was that the first "cosmonette" landed only a few hours before the Hawk (that is, Bykovsky) brought his cosmic marathon (it had lasted nearly 71 hours) to a safe conclusion.

On June 14, 1963, the day that Bykovsky sets off into space, Valentina and Gagarin see him off. Two full days go by. Then Bykovsky, through his headphones, hears a new voice in space, a woman's voice. "Separation from the carrier rocket completed," Valentina tells earth, while Bykovsky listens intently. "It is beautiful up here. I can see the horizon; what gorgeous colours. . . ."

For a while the cosmonaut lets her give vent to her emotions. He knows that she, like himself, has lived for this moment for more than ten months—for that is how long the special training for their tandem flight lasted.

Bykovsky waits until Valentina stops speaking; then he calls her up:

"Tsyaika, Tsyaika, Valeri here. Congratulations on your

jump into space. . . . " Valentina replies immediately: "Valeri, Valeri, I receive you perfectly. It's just as if you are sitting here beside me."

"Of course I am sitting beside you," Bykovsky exclaims. "We are travelling through space side by side. . . . "

Their excited conversation continues. Cosmonaut and cosmonette forget to use their call signs. They are no longer thinking of the earth, where every word spoken far out in space is being heard.

That same evening,* when the time comes for them both to go to sleep, Valentina suggests that they sing a song together. She has finished eating, and her colleague is just finishing his meal, packed partly in tubes, partly in plastic bags. He opens his mouth to start the first verse, when he suddenly sees some pieces of cheese floating on his breath through the cabin. He bursts out laughing and tells Valentina, who is wondering what's going on, what has happened.

On one occasion Bykovsky is worried about his travelling companion. Both he and mission control have called her up repeatedly after a rest period. At last, after several minutes which seem like hours, she answers and confesses that she . . . has overslept.

"It won't happen again," she assures Sergei Korolyov who has come to the microphone in person to find out what is happening. Immediately afterwards Valentina, as usual, passes on information about temperature, pressure and humidity in the cabin.

Vostok-5 lands on June 19, at six minutes past two in the afternoon, almost three hours after Valentina has returned to earth. The landing sites are nearly 800 km apart.

Hardly has Valentina, ejected from the descending capsule at a height of 7 km, landed in her scarlet spacesuit, when a crowd of enthusiastic people converge on the site. Members of the recovery team do their best to keep dozens of excited children away from the space capsule, which has landed close by. Meanwhile Valentina is busy shaking hands, but as soon as she gets an opportunity, she gets out of her clumsy and unbecoming spacesuit, helped by a fellow-cosmonette.

Six days after their return to the Baikonur base Valentina and Col. Bykovsky were welcomed at Vnukovo Airport,

*Evening in Moscow. For a spaceship circling close to earth day and night alternate every ninety minutes.

Moscow, by Prime Minister Khrushchev. Firmly Valentina marched at her malc colleague's side along the red carpet which stretched from the plane to the reception rostrum, where the Prime Minister and other government dignitaries were waiting. Bykovsky was in uniform; Valentina was wearing a dark grey suit. Hundreds of thousands of excited Russians lined "Cosmonaut Avenue", the route between Vnukovo and Moscow, cheering till they were hoarse. At the head of the procession came Khrushchev, Valentina and Valeri in a black limousine, followed by a second car carrying the cosmonauts Nikolayev, Titov, Popovich and the first traveller in space Gagarin. Three days after the official tribute in Red Square and a well attended reception in the Kremlin, where, apart from chief designer Korolyov, ten cosmonauts in training and four candidate cosmonettes were present, it was the turn of the journalists.

There was not a single empty seat at the press conference, which was presided over by Mstislav Keldysh, president of the Academy of Sciences. In the course of the interviews Valentina declared frankly that if she married, she would continue her spaceflights. Some of the reporters immediately pricked up their ears. Was she already in love?

"I need have no fear that I'll be a spinster," she replied evasively.

What had she thought about most during her flight? Valentina blushed. "My mother," she said.

The romance between Andrian Nikolayev and Valentina, long suspected, later proved true: on November 3, 1963, the two were married.

The following year, on June 8, the first "space child", Yelena, was born. Not only family and friends were interested in the attractive dark-haired baby. Both her father and her mother had spent a considerable number of hours in space. Could this be detrimental to their progeny?

In this connection it was not weightlessness and its possible effects which was the first consideration. The cosmic radiation factor, which on prolonged flights in space without adequate protection undoubtedly has adverse effects, caused greater concern. Valentina's orbit had been between 181 and 231 km, her husband's between 181 and 235 km. At that height cosmonauts are subjected to a certain measure of primary cosmic radiation.

MANNED SPACEFLIGHTS IN THE FRAMEWORK OF THE VOSTOK PROJECT

Launching date	Apr. 12, 1961	Aug. 6, 1961	Aug. 11, 1962	Aug. 12, 1962	June 14, 1963	June 16, 1963
Name of spacecraft	Vostok-1	Vostok-2	Vostok-3	Vostok-4	Vostok-5	Vostok-6
Crew	Yuri Gagarin	Herman Titov	Andrian Nikolayev	Pavel Popovich	Valeri Bykovsky	Valentina Tereshkova
Weight in kg	4725	4731	4722	4728	4720	4713
Initial orbit Min·max. height	131 327	183 244	181 235	180 237	175 222	181 231
Orbiting time in mins.	89·1	88·5	88·3	88·4	88·3	88·3
Angle with equator in degrees	64·95	64·93	64·98	64·95	64·97	65·00
Number of orbits	1	17	64	48	81	48
Duration of flight	1 h 48 min.	25 h 18 min.	94 h 22 min.	70 h 57 min.	119 h 6 min.	70 h 50 min.
Landed on	Apr. 12, 1961	Aug. 7, 1961	Aug. 15, 1962	Aug. 15, 1962	June 19, 1963	June 19, 1963

The Russian professor Alexander Neyfach has carried out a great number of experiments to determine the effects of cosmic radiation and intensive x-rays on the embryo. Both male and female cells were subjected to such radiation. It appeared that the nucleus of the cell—which contains *inter alia* the chromosomes (carriers of hereditary characteristics)—is particularly sensitive to this form of intensive radiation.

The remarkable discovery was made that a cell continues to develop for some time after the nucleus has been killed by radiation. It was not until later that Neyfach discovered that, particularly in the early stages of an embryo's development, the nucleus passes instructions to the rest of the cell for six to eight hours at a time, and that these instructions determine the subsequent development for that length of time. If after this period the nucleus does not become active again (as after intensive radiation), the entire cell dies. This kind of problem might be encountered if a pregnant cosmonette were sent into space. The Nikolayevs' case, however, was different. They had spent respectively 100 and 75 hours in space and were subjected to primary cosmic radiation. The question which the scientists asked themselves was whether this radiation would prove to be strong enough to affect the genes, the carriers of hereditary traits, in both male and female chromosomes. The condition of baby Yelena, who has since grown into a healthy little girl, was reassuring on this point.

After the birth of their little daughter, both Andrian and Valentina studied aircraft engineering at the Shukovski University, a military academy, and at the same time continued their cosmonaut training.

As is customary in Russia, *babushka* (grandmother) looked after the baby when father and mother were away. . . .

THREE IN SPACE

Voskhod: first three-man capsule – Cosmonauts in mufti – Doctor in space – Soft landing – Interim design

After each successful Soviet space marathon, the so-called "experts" lose no time in predicting what the next flight will bring. Such was the case after Valentina's flight which later proved to have been the last in the successful Vostok series. One conjecture followed another. It stood to reason that the Russians would now be expected to plan a two-man spaceship. The Americans were preparing their Gemini two-seater capsule, and it seemed improbable that the Russians would allow themselves to be beaten in this project. It was known that the first manned Gemini was to be launched early in 1965, so it might be expected that the Russians would attempt to beat their rivals in 1964. This, in fact, is what happened, albeit much later in the year than was anticipated. However, it was a complete surprise when the Russians launched not two, but three cosmonauts in one spaceship. "Those Soviets always do the unexpected," the American spaceman Scott Carpenter complained, thus doubtless voicing the feelings of a great many experts in his country.

The three Russian cosmonauts, Komarov, Feokistov and Yegorov, also had a surprise. While they were, as one might say "seen off" by Nikita Khrushchev when they departed on October 14, they were welcomed on their return the next day by Brezhnev and Kosygin. Nikita Sergeyevich had meanwhile had to retire from the scene.

No spacesuits

Several weeks before the start of Voskhod-1 a large group of cosmonauts had flown from "Star Village" near Moscow to Baikonur. It was not known which of them would be sent up into space. Not until shortly before the new spaceship was launched did the State Commission which attends every manned take-off decide that Vladimir Komarov (37), Konstantin Feokistov (38) and Boris Yegorov (27) were to be the first trio in space. At the same time stand-ins were named to take over in case of an emergency.

Vladimir Komarov belonged to the first group of cosmonauts from among whom in 1961 Gagarin had been picked to be the world's first space traveller. Feokistov and Yegorov had joined the ranks much later; they had undergone a relatively short training. They were essentially scientists, the one an engineer, the other a doctor. The control and command of the spaceship were entrusted to Vladimir Komarov, a cosmonaut of the classical school: an experienced fighter pilot and technically knowledgeable.

October 11

On the eve of the launching, chief designer Sergei Korolyov as usual drops in on "his" cosmonauts. He wants to know how they are feeling, what the doctors have said. Then he excuses himself—he has something else to do. "Make sure you have a good rest," he says.

Next morning the special cosmonaut coach is waiting when the sun's rays are only just touching the wintry steppe. Komarov, Feokistov and Yegorov come out and take their seats. It is remarkable that they are not wearing spacesuits, but only light training garments, consisting of a blue jacket and grey trousers. On their heads they wear small helmets equipped with earphones and microphones. The Russians have such faith in the Voskhod's construction that on this occasion the heavy, clumsy spacesuits are not worn, although they are carried on board in case of an emergency. The spaceship could, for instance, be hit by a meteorite, which might cause the air to escape from the capsule—slowly, one hopes. There are no ejector seats—they would take up too much room. Whereas on Vostok flights the cosmonauts were catapulted from their cabin at a height of seven kilometres, the Voskhod trio will land in their capsule. Special retro-rockets will reduce the landing speed to practically zero.

Start

The three cosmonauts take leave of the members of the State Commission and of their colleagues. After a last warm embrace—an ancient Russian custom—from cosmonaut No. 1, Yuri Gagarin, they climb the fourteen steps up to the lift, one after the other. At 10.30 exactly, the first-stage engines of the multi-stage RNV rocket are fired.

Tass announces:

"On October 12, at 10.30 Moscow time, the first Voskhod spaceship, carrying a crew of three, was put into orbit round the earth by a new, powerful carrier rocket.

"On board are the citizens of the Soviet Union Pilot-Cosmonaut Vladimir Komarov, Colonel of Engineers and commander of the spaceship; and Boris Yegorov, doctor of medicine and Konstantin Feokistov, Master of Engineering Science.

"This new flight in the cosmos aims at the following objectives:

—to test the new guided multi-seater spaceship;
—to check the capacity for work and the interaction during the flight of a group of cosmonauts consisting of specialists in different fields of science and technology;
—to carry out scientific physical and technical investigations in the conditions of spaceflight;
—to continue the study of the effects of various factors of spaceflight on the human organism;
—to carry out extensive medical and biological research in conditions of prolonged flight.

These investigations are being carried out with the help of instruments on board the spaceship and with the direct participation of a scientific worker and a space doctor.

"The Voskhod has been put into an orbit close to the prescribed one. According to preliminary information, the period of revolution round the earth is 90 mins. The minimum and maximum distances from the earth are respectively 178 and 408 km. The angle of the orbit with the equator is about 65 degrees. Two-way radio communications are being maintained with the spaceship.

"The crew have reported that they have stood up well to the launching and the transition to weightlessness. They are in good condition."

21. (*Above*) Valeri Sevastyanov in the work compartment of Soyuz-9.

(*Below*) The command module of Soyuz-9 has landed safely after its marathon flight. A member of the recovery team has climbed on top of the capsule to help the crew get out.

22. (*Above*) A photograph of the moon's surface, taken by Luna-9. In the foreground a section of one of the corolla antennae is visible.
(*Centre*) The '*Kosmonaut Vladimir Komarov*', used for tracking and guiding spacecraft.

(*Below*) Large parabolic antennae at the Centre for Remote Space Communications in the Crimea.

As soon as the news had been announced and broadcast over the loudspeakers, large numbers of Muscovites as usual gathered in Red Square. To them, too, the flight came as a complete surprise, even though there had been rumours that a sensational experiment was being planned. What to think of this flight?

The fact that there were three cosmonauts on board, and not just one, was sensational enough in itself, but nevertheless all sorts of conjectures were raised concerning possible further plans which might be inferred from the rather vague Tass phraseology.

The lack of actual information inspired many western correspondents in the Russian capital to flights of fancy. The suggestion most often heard was that a second spacecraft would soon be launched to link up with the first.

But this was by no means on the programme. There was no question of a rendezvous being prepared. The space-trio began by comfortably giving a television show and sending messages to the countries which their capsule overflew. The cosmonauts transmitted warm greetings to participants in the Olympic games in Tokyo and had a pleasant chat with Khrushchev and Anastas Mikoyan. "The flight in the cosmos continues," Tass reported. "The unforgettable event is followed with suspense by those on earth."

Doctor in space

During the flight each member of the crew had his own special tasks.

Vladimir Komarov was, first and foremost, entrusted with the direction control of the spaceship, which weighed 5320 kg. He also had the task of testing the ion propulsion devices attached to the outside of the capsule to make it possible to turn the craft in any desired direction.

On earlier flights this had been done solely by means of small valves. By allowing a small quantity of gas to escape, the craft turned in the opposite direction, in accordance with Newton's law that "action involved reaction". The Voskhod had this "classical" installation as well, but in addition the ion propulsion devices had been mounted for the first time to find out how they would work in cosmic space. This type of engine is based on the acceleration of electrically charged particles (for instance caesium-ions) in a magnetic field. These small engines are not very powerful, but little energy is required to make a spaceship revolve on one of its axes, that is, its centre of gravity. This was to be only too obvious during the flight of Voskhod-2. On this occasion the capsule began to turn when Leonov, on his walk in space, pushed himself off from the outside.

Naturally it is an entirely different matter when a spacecraft is to change direction, that is, to move the centre of gravity out of the original trajectory. This requires exceptionally powerful rocket engines, particularly if the course correction is to be considerable.

Konstantin Feokistov, the scientist on board Voskhod, is closely involved in the development of new spacecraft, so it is not surprising that his chief task was to check all the apparatus carried. He also carried out scientific observations. Among other things he observed the polar lights which during the flight of the "Ascent"* happened to be particularly magnificent, "the most impressive phenomenon we saw during our flight," said Feokistov. He moreover measured the height above the horizon of various brilliant stars, using the classical type of sextant which is also used at sea. The chief purpose of this operation was to determine whether the horizon of the earth, as seen from space, was sufficiently sharply defined to be used as a point of departure for "shooting" the stars. Feokistov came to a very definite conclusion. He was of the opinion that in future cosmonauts would be able to use a sextant to fix their course. (This was done on the Soyuz flights.)

Of course the first responsibility of the youthful looking doctor, Boris Yegorov, was to keep an eye on the health of the cosmonauts, including his own. Although still young, he had been occupying himself for many years with man's balancing organism in the inner ear, which plays so important a part in the reaction of a human being to weightlessness. Yegorov carried out many experiments in connection with this vestibular mechanism.

In many respects his position on board the spaceship was unique. For the first time in the era of space travel the medical men did not need to rely solely on the information received

*"Voskhod" means "ascent".

by means of the strips of graph paper which, inscribed by sensitive pens, rolled out of the recorders in the co-ordination centre. This time they received first hand information.

Immediately after the start, when the enormous forces of acceleration prevented the cosmonauts from using their arms and legs, Yegorov had asked his comrades how they felt, and from that moment onwards he had kept a constant eye on them. Thus he could, on the return of the first three-man spaceship, give an accurate report of the remarkable symptoms of space-sickness which he himself, and Feokistov in particular, had experienced during their 24 hours of weightlessness.

At the press conference following the flight of Voskhod-1, Yegorov hardly referred to this point. He only mentioned that he and Feokistov had had the sensation that they were flying upside down when they closed their eyes. Vladimir Komarov, who was a fighter pilot and whose cosmonaut training had been particularly thorough, had not experienced these unpleasant sensations at all.

In the course of the flight of this first cosmic three-seater it was not only the human beings who were observed from a bio-medical point of view. The capsule also carried drosophila flies (banana fruit-flies) and plants, chiefly for the purpose of observing the effect of weightlessness on these forms of life. Drosophila flies are particularly suitable for this kind of research, since they multiply rapidly, so that the effect of certain factors (such as, in this case, weightlessness and radiation) on successive generations can be observed within a short time.

Landing

More than 24 hours after lift-off, Voskhod-1 begins its preparations for landing. First of all the spaceship takes its bearings by the sun by means of sensitive sensors, and then stabilises itself. At that moment the main retro-rocket (the spacecraft carries a reserve retro-rocket as well) is correctly aligned. At a command from earth the retro-rocket is fired and Voskhod's speed diminishes. The compartment containing the instruments and the rockets are disconnected and the spherical capsule with the three cosmonauts races towards the atmosphere. Down below, in the flight control centre near the launching base, tension reaches unknown heights, for this will be the first time that cosmonauts will land in their capsule. Will Voskhod's parachute function properly? And the small retro-rockets . . . will they fire at the correct moment to prevent the capsule crashing into the ground, out there in the steppe?

During the descent contact with the capsule is temporarily lost. A layer of air, electrically charged as a result of friction, surrounds the spacecraft, so that radio-waves cannot reach it. But then, after minutes which seem like hours, a message is received from the recovery teams who are flying in planes and helicopters to the predetermined landing area near Kustanai: "The landing of the capsule has been noted". Chief designer Sergei Korolyov immediately takes the microphone: "We are waiting for the results of the landing. How is the crew?" Another tense minute of waiting. Then comes the reply in the remarkable, impersonal style of all reports: "Three people have been observed on the landing site."

Gagarin, who is standing next to Korolyov, wipes the sweat off his forehead. Komarov, Feokistov and Yegorov are back on earth!

Interim spaceship

On October 21 follows the by now traditional press conference in the crowded auditorium of the University of Moscow, situated on the Lenin Hills. Two thousand journalists from foreign as well as from Russian papers are gathered here to bombard space trio with questions.

"How was the landing?" one of the correspondents wants to know.

Komarov: "The landing of the spaceship was softer than the stopping of a modern lift."

A rather bold question: "Could you land on the moon in this spaceship?"

"It has not been designed for that purpose," says Komarov, but, as is to be expected, he does not expand further on the question. All replies are exceptionally brief and non-committal. The Soviets are very careful not to give too much information about their new space ace. It is, for instance, not made clear whether the Voskhod is an entirely new type of spacecraft, perhaps with greater capacities, or in fact no more than an amended Vostok.

Komarov only says: "It is a new spaceship with room for three people." That is all.

Subsequently it became known that, although the Voskhod contains a number of new constructional details, it is nevertheless a direct derivation of the one-man spacecraft, in fact an adaptation of the Vostok. The removal of the large ejector-seat created enough space for a three-man crew. When I asked my guide, an engineer, in Moscow, why I was allowed to see the Vostok and the Soyuz, but not the Voskhod, he said frankly: "We prefer not to show it. It was only an experimental spaceship in which a number of new techniques were elaborated." Voskhod clearly was an interim version.

Towards the end of the press conference Mstislav Keldysh, the president of the Academy of Sciences, replied to a number of questions, some of which were really quite tiresome. One journalist, for instance, inquired naïvely about the composition of the material which formed the Voskhod's heat shield. Naturally Keldysh refused to tell him and added that this was "one of the secrets of Soviet experts".

He was, however, prepared to talk about the general significance of the mission. Voskhod's flight, he said, was a first step towards establishing a large space station in orbit round the earth, but it was also of importance in connection with a manned flight to the moon. The order in which Keldysh mentioned these two aspects is of great significance. As had appeared previously, it was here confirmed that the Soviets thought a manned, permanent space station in orbit round the earth to be of greater importance than a flight to the moon.

Summary of the Voskhod-1 flight

Launched on October 12, 1964 at 9.30 a.m. Moscow time. Weight 5320 kg. Transmitter frequencies: 17·365, 18·035, 19·9944 and 143·625 megahertz. Angle of trajectory with the equator: 65 degrees. Orbiting time: (90·1 minutes. Perigee: 178 km; apogee: 408 km. Duration of flight: 24 hrs. 17 mins. Number of orbits: 16. Landed on October 13, 1964, 9·47 Moscow time.

GIANT STEPS IN THE INFINITE

Leonov: first space walker – Mechanical training – Air-lock chamber – Intensive training – The tired "walker" – Perilous ending

Leonov, first walker in space

"The hatch is open. I am getting out."

These everyday words were spoken on March 18, 1965, by a man who at that time was 500 km from earth.

When he put his helmeted head outside the narrow hatch opening, he looked down into a bottomless well.

Yet he did not hesitate for a second. Entirely at ease, he pushed his shoulders and arms through the opening and looked around with interest. Below him he saw the sharply defined contours of the deep blue Black Sea. With one glance he surveyed the world from Africa to the Ural. He saw the massive mountain ranges of the Caucasus, snowcapped here and there. Above, myriads of brilliant stars hung in the velvety, dark, lifeless depths of space. And between, like an unusually strong arc-light: the sun, silvery-white. The man could feel the heat of its scorching rays even through the thick filterglass of his visors.

Carefully holding on to the edge of the hatch, he pushed himself further out into the unknown. He did a few simple exercises, like an embryo gymnast, and then pushed himself off. He floated slowly into space and a thin "umbilical cord" unfolded between him and his capsule.

At that instant a nearby voice, trembling with emotion, called out: "At this moment a man is floating free in space." Radio signals sped the message to the waiting earth down below. A few minutes later loudspeakers, positioned along the beach of the Black Sea, told the holidaymakers the news: "Alexei Leonov has left the cabin of his spaceship Voskhod-2. Pavel Belyayev, who has remained inside, reports that the cosmonaut is in excellent condition."

Involuntarily, thousands of people looked up, but of

course there was nothing to be seen. However, thanks to modern science, millions of television viewers were able to witness the event. A camera mounted on the outside of the capsule transmitted pictures of the cosmonaut hastening towards Siberia with giant strides, at 28,000 km per hour. Ten minutes later he was inside the capsule once more and the most important experiment of the 24-hour flight was at an end.

Next day Alexei and Pavel landed their capsule somewhere near Perm. Once again the Russians had achieved an almost incredible scoop.

Mechanical training

Months before the world was surprised by this flight, preparations for it had begun in all secrecy.

Why should such an experiment, at first sight so daring, be undertaken at all? For propaganda purposes, as some malicious people thought? No, the men responsible for Russia's manned spaceflights had their good reasons. We have mentioned them earlier in this book. First of all there was the fact that some day—that is, within a few years—enormous space stations would have to be established in space, if truly fruitful manned spaceflights were ever to be achieved. Space stations with a large staff, mainly consisting of scientists who could keep an "all-seeing" eye on the development of the weather pattern. Space stations for carrying out all manner of scientific experiment in which zero-gravity and cosmic radiation are essential requirements.

Space stations, possibly, for curing patients to whom weightlessness might bring considerable relief. And—last but not least—space stations to provide a springboard towards the moon and the other planets.

Such gigantic space-ports, possibly tens or even hundreds of metres in diameter, could of course not be built in one go. They would have to be sent up in the form of components, to be carefully assembled in space. Viewed in this light, the significance of Leonov's walk-out was immediately apparent.

His excursion, however short, was of immense importance not only to technicians, but also to space doctors, for Leonov found himself in a totally new situation, never before experienced by man.

There was first of all the lack of any point of reference. The cosmonauts and their Voskhod were weightless as they flew round the earth at a speed of 28,000 km per hour. Nevertheless, on stepping outside, Leonov continued to float beside the capsule, for he had the same speed as the spaceship, a speed which was not slowed down by anything at all. At a height of several hundred kilometres there is practically no air, so there can be no question of any "headwind".

Leonov was attached to the spaceship by a lifeline five

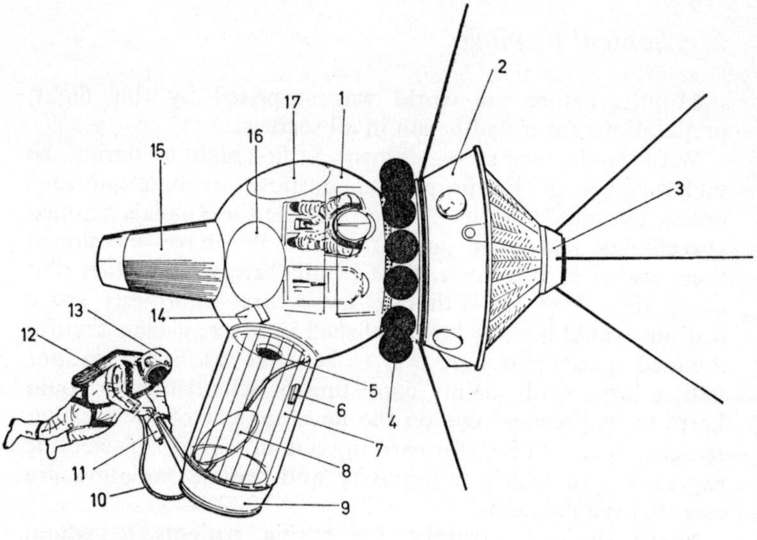

Fig. 13. Voskhod-2.

1. Spherical two-seater command module (the right-hand side occupied by the flight commander).
2. Instrument compartment.
3. Main retro-rocket.
4. Pressure bottles of gas-jet stabiliser system.
5. Optical orientation apparatus Vzor, mounted in hatch.
6. Service panel in air-lock chamber.
7. Inflatable air-lock chamber.
8. Outer hatch (opening inwards).
9. Metal collar of air-lock chamber.
10. "Umbilical cord" with telephone cable.
11. Detachable film camera.
12. Cosmonaut's "life-pack".
13. Helmet with inbuilt sun visors.
14. Instrument panel.
15. Reserve retro-rocket.
16. Parachute hatch.
17. Entry hatch.

metres in length—the so-called "umbilical cord", and this was by no means an unnecessary luxury. For what would have happened if there had been no such lifeline, if the cosmonaut should accidentally have pushed himself off against the spacecraft? He would have floated irretrievably away, without any chance of ever returning to the capsule. There is no friction in space and swinging one's arms and legs results only in uncontrolled revolutions on one's own centre of gravity.

In addition there was the problem of orientation. In a weightless condition, and especially when there is no floor or ceiling, it is very difficult to orientate oneself in space. "Above" and "below" have lost all meaning.

A further danger was the lack of air pressure. If Leonov's spacesuit were to have leaked, the pressure inside his lungs would no longer have been counteracted by outside pressure, and he would literally have exploded. Fortunately the scientists did not have to worry too much about the dangers of radiation, heat or cold in their preparations for the experiment. They had impressed upon Leonov that he should on no account stay outside in space for more than ten minutes. Voskhod-2's chosen trajectory, moreover, lay well within the dangerous Van Allen Belts. For safety's sake a number of Kosmos satellites had, in addition, been launched shortly before the start of Voskhod-2, to provide accurate information concerning the radiation at the altitudes which the new spaceship was to reach.

Air-lock chamber

When the news of Leonov's sensational walk in space first became known, it seemed as if Voskhod-2 was an entirely new type of spaceship, but this later proved not to be the case. In any event, the name alone might make one suspect that the two Voskhods were technically very similar. The chief difference between the two (and this was apparent from the first) was a uniquely constructed air-lock chamber. *Why* it was unique was initially kept a secret. As usual, Tass was extremely reticent about technical details, which would have been, to say the least, of great interest—especially to the Americans.

The American Gemini space capsules had to be depressurised before an astronaut could get out, so that the man who stayed behind was subjected to practically the same dangers as his colleague outside.

But Voskhod-2 had an air-lock! Thanks to this apparatus Leonov could leave the spaceship without the necessity of pumping the air from the capsule, where commander Belyayev remained during the space walk. Here, normal atmospheric pressure was maintained, consisting of a mixture of nitrogen and oxygen, as we have on earth.

The Soviet technicians had opted for a simple, but nevertheless effective solution, which enabled them, with minimal alterations, to adapt an existing Voskhod for a cosmic excursion, whereby the cabin need not be depressurised. The construction of the air-lock chamber was as ingenious as it was simple: a collapsible tunnel was attached to the exterior of one of the hatches. In the spherical Voskhod-2 capsule, next to the seat of the second pilot, there is a removable, circular hatch in which the optical orientation apparatus, the Vzor, is mounted.

The inflatable air-lock chamber was mounted around the outside of this hatch. Even folded like an accordion—as it was on launching—the air-lock took up so much space that an indentation had to be made in the nose-cone which protects the spacecraft while it traverses the atmosphere.

When Voskhod-2 had been put into orbit and the protective nose-cone had been jettisoned, the air-lock chamber, which was made of a man-made material, was inflated. Because of the fact that it had to be absolutely airtight, the outer hatch had been so constructed that it opened towards the inside.

After the first orbit had been completed, the air-lock as well as Leonov's special spacesuit were thoroughly examined. Then Leonov opened the cabin-hatch and floated into the air-lock, which had meanwhile been pressurised. After the space-walk, this chamber must have been jettisoned (although the Russians have never confirmed this), as it could not have survived re-entry through the atmosphere, when the exterior temperature of the capsule at times exceeds a thousand degrees Celsius. Moreover, the chamber would have made it impossible to use the Vzor which is switched on for the return to earth.

Intensive training

"Between April 1964 and March 1965 I cycled about a thousand kilometres and ran hundreds of kilometres cross-country," Leonov told reporters who were questioning him about his training for the flight.

But of course Leonov and Belyayev had not only to undergo physical training. During the final year of preparation in particular, they took part in the construction of their spaceship. And when, after numerous small amendments, a definitive design had at last been achieved, they started on the final phase: minute preparations of every small detail of the remarkable journey they were about to make.

"As Voskhod-2 differed in many respects from previous spaceships," Belyayev said later, "and the project moreover involved a man leaving the capsule, our training programme was very complicated. We devoted a great deal of time to studying the construction of Voskhod-2 and its equipment and took part in testing the apparatus. We also used a flight simulator in which we mastered all the details of the voyage. In this simulator Leonov and I learned to co-ordinate our actions during the various phases of the flight until our reactions became automatic, especially as regards Leonov's walk in space and his return to the spacecraft. A great deal of attention was devoted to this part of the flight. I had to orientate the capsule, attend to the air-lock chamber, read the instruments and carry out other experiments in connection with Leonov's excursion in space."

All this training took place in a gigantic vacuum chamber, simulating the actual flight. The only thing lacking was weightlessness. . . .

And even this was imitated, if only for short spells at a time. The entire cabin was mounted in a Tupolev-104, a twin-engined jet-plane of the type used on passenger routes, for instance between Amsterdam and Moscow. By making this plane describe a parabolic trajectory at a high altitude, all the objects on board became weightless. While the plane, like a brick thrown upwards, continued to ascend for a little while before it fell back towards earth, Leonov hurriedly buckled on his equipment and floated into the air-lock. At that moment—after about half a minute—gravity returned: the pilot had put the plane once more on a horizontal course. A second parabola, however, gave Leonov sufficient time to open the outside hatch and float into the hold of the aircraft, which had been padded with thick mattresses. This lining was necessary, for the cosmonaut was weightless and when the engines were restarted he might have been floating upside down.

Thanks to the Tupolev, the candidate space-walker became quite adept at the technique. Later he was to say: "Everything really went just as during training."

But the preparations did not end here. Dr. Yevgeni, the space doctor, wanted to make sure that the cosmonaut would be physically equal to the experiment which awaited him. He would be the first man to cross a new threshhold, both literally and figuratively. Would he be able to face all the unaccustomed factors at once: the immeasurable emptiness, weightlessness, the lack of any point of reference?

The doctors put Leonov for a whole month in a soundproof room. There was no contact with the outside world. The doctors could observe the solitary guinea-pig day and night, but he could not see them. During that month in isolation Leonov followed a strict régime. Sometimes he made drawings —he is quite a good artist. Later he was to record the overwhelming impressions received in space in both oils and watercolours.

After this month the cosmonaut was not allowed time for readjustment. He was dressed in flying kit and taken straight to the airfield, where a Mig-15 was ready for take-off. Leonov took the controls; a flying instructor sat behind him. A short time afterwards the Mig was zooming through the azure sky, faultlessly looping the loop, finally, after several further successful stunts, to make a perfect landing. The doctors concluded that for Leonov no psychological barriers existed. But training continued. The cosmonaut's sense of balance was subjected to trial in special cabins revolving and whirling in all directions. This organ, which is situated in the middle ear, was to be greatly tested in space. Leonov had to learn to orientate himself by means of his eyesight only. Meanwhile he made one parachute jump after another, keeping his parachute closed as long as possible, for a parachutist in free fall is weightless, apart from the resistance provided by the air. Of course, Pavel Belyayev was not spared this type of exercise, for in an emergency he might also have to leave the capsule.

The tired "walker"

Back in the cabin, Belyayev could keep an eye on his colleague by means of a television camera. The presence of the commander was by no means superfluous. During the preparations for the walk-out, soon after launching, he assisted Leonov in testing his equipment. For over an hour before his egress, Leonov, instead of breathing the normal air within the cabin (consisting of nitrogen and oxygen), used his personal supply of pure oxygen. This was essential, as all traces of nitrogen had to be removed from his blood, for outside in space Leonov would experience much lower pressure, as a result of which the nitrogen in the bloodstream would form bubbles, which might be fatal if they reached the heart. This phenomenon, which had claimed a large number of victims before its nature was understood, is known to divers as "the bends".

Hardly had Belyayev given the order, when Leonov got into the air-lock chamber. The commander closed the internal hatch and the air was slowly pumped out of the chamber, while Leonov checked his spacesuit, his helmet, his gloves, the "umbilical cord" and his oxygen apparatus. Everything was in order. Seconds later the outside hatch opened. The pathway to space lay open.

But Leonov had to wait for a signal from Belyayev before getting out. The commander sympathised with his colleague's impatience, but nevertheless wanted to adhere exactly to the programme. The moment at which the walk-out was to begin was closely linked with the position of the spacecraft at that time: above the Black Sea. This would make a direct television transmission to Moscow and its surrounding area possible. A special mounting on the rim of the outside hatch held a film camera which was to record the walk-out in colour. Leonov was to take this camera with him when he re-entered the spacecraft.

Five hundred kilometres below Voskhod-2, television viewers witnessed the experiment. The pictures were not very sharp, but nevertheless one could clearly see the cosmonaut's helmeted head coming out. Once his whole body was outside, Leonov pushed himself off and floated away until the cable linking him to the capsule was fully extended.

But let Alexei speak for himself:

"While I was floating in space, over the Kuban steppes, I was in radio contact with Yuri Gagarin. He asked me anxiously how I was feeling. I replied that I could see an enormous amount, but that it was difficult for me to describe it all. Yuri laughed and then asked me about my morale. I again said: 'I can see such a lot. . . . ' I was completely fascinated by my impressions. Yuri gave up. He realised that I was completely engrossed. We are old friends; half a word is enough for us."

Meanwhile both the capsule and the cosmonaut were moving at a speed of 28 000 km per hour, but this had no meaning for Leonov. "Up there I had no idea at all of the enormous speed at which the Voskhod and I were moving in space. When I looked at the capsule it appeared to be hanging immobile above the cosmic abyss. On earth one experiences the sense of speed because of the trees and the houses which race by. Even with one's eyes closed one is aware of speed, because of the sound of the engine and the bumping of the car. In space there is nothing like this. The spacecraft made no sound. All I could hear was Pavel's voice via the telephone link. To my mind the capsule was a planet like any other—the only one in the immeasurable ocean of space in which we found ourselves. The spaceship appeared enormous to me, and that, of course, was because there was nothing to compare it with.

A fantastic sight! The stars appeared to be motionless. The sun seemed as if sewn onto black velvet. In this universe only one thing moved and that was earth. We ourselves seemed not to be moving at all. I only felt a slight movement in the spaceship at the moment when I got out. When I pushed myself off, it moved for an instant in the opposite direction."

Leonov had great difficulty in salvaging the film camera, and also he appeared to be rather tired.

"My right hand was holding several objects, and it was therefore none too easy to re-enter the air-lock chamber. Now I noticed that, after making various movements in space —for instance, moving my head—I had become rather tired. It is not all that easy to work in a spacesuit. Every movement requires greater exertion than on earth. Moreover, in a spacesuit one is aware of carrying one's own air pressure, a feeling unknown to us on earth. To give an example: someone on earth wearing gloves does not feel their pressure, but in space one does, and that pressure has to be counteracted by a measure of exertion. This explains why I was somewhat tired after ten minutes out in space. When it was time for me to go back, I started to put everything I had with me into the air-lock and then began to get in myself. Just as I was about to do so, I saw the camera floating beside the capsule. I couldn't help laughing. I got hold of it and put it back into the chamber. No use: it immediately floated up and away again. Drastic measures were indicated. I again got hold of the camera, put it in the chamber and held it down with my foot. Then I got in myself. I immediately looked to make sure that the camera was still there. And sure enough, it had bowed to the inevitable. My struggle with the camera had tired me even more. I was sweating like a pig, something which rarely happens to me. Even during the prolonged exercises which preceded the flight I hardly ever perspired. Obviously my exertions had been particularly great.

Now I quickly closed the outer hatch and pressurised the chamber. When the pressure was normal, the inner hatch opened and I was beside Pavel in the cabin once more. He greeted me with a beaming smile: 'You are a champion, Lecha!' "

Perilous ending

Leonov had been out in space for exactly ten minutes. The flight continued. The cosmonauts made the necessary observations, they checked each other's blood pressure, breathing rhythm and heartbeat, and of course Leonov could not resist the temptation of seizing his drawing pad and pencil to try and record some of his impressions. He also brought the *Bortovoi Zhurnal* up to date, in which all experiences were set down in the smallest detail. There was also a small tape-recorder, which recorded all conversations with the earth. Then, 24 hours after the start, the moment of landing approached and with it came problems.

All previous Russian spaceships had landed automatically, on a signal from earth. This meant that all through the complex landing operations the cosmonauts could relax. Automatic apparatus orientated the spaceship to the sun, so that the retro-rockets were correctly aligned. Once the correct position had been reached, the retro-rockets fired automatically, slowing down the spaceship to a speed in which it could not stay in orbit. The compartment housing the retro-rockets was jettisoned, after which the spherical capsule raced towards the earth's atmosphere, in order, approximately half an hour after the start of the descent, to land by means of a large parachute "within a predetermined area of the Soviet Union".

But this time it happened differently. When during the seventeenth orbit the spaceship received the signal to orientate itself by the sun, it did not respond correctly. Immediately Belyayev asked Baikonur for permission to make a manual landing.

"You can imagine," he said later, "that we were really quite excited when we saw that the sun-orientation system had failed just as we were getting ready to land on the automatic pilot. At last we had an opportunity to make a manual landing and to provide further proof of the exceptional qualities of our cosmic machine, now truly guided by human hand. There was only one thing we were afraid of: that permission would be refused. We might have been instructed to complete another orbit and then to land on the automatic pilot.

Thirty seconds passed before a decision was taken, based on our report and our request, and they were a long thirty seconds. But we received permission to make a manual landing in the eighteenth orbit."

We may doubt whether this cheerful report corresponds entirely to the truth. There are indications that the mood during these critical moments was not exactly happy. Moreover it seems unlikely that in the next orbit the automatic orientation system would suddenly have functioned correctly once more. Mission Control simply had no choice, so it did not take long to make a decision.

On the other hand, the cosmonauts had been thoroughly trained in making manual landings, and the spaceship was fully equipped for this operation.

"As an ex-fighter pilot I had landed a great many modern fighter planes," said Belyayev, "but Voskhod-2's speed was many times greater. I had to land this super-fast craft completely manually in the predetermined area. It must be obvious to everyone that this involved a great deal of responsibility. The slightest deviation in direction, or the wrong decision concerning the firing of the retro-rockets would at best result in a landing in the wrong place, and at worst in no landing at all, merely a change of position in space.

23. Photograph of the entire earth taken by Zond-7 while flying at a distance of 2000 km above the moon. Because of the unfavourable position of the sun, the surface of the moon shows little detail.

24. (*Above*) In the Centre for Remote Space Communications in the Crimea.

(*Below*) Bombay: Loading the Zond-5 capsule (in a container), which came down in the Indian Ocean. The capsule is practically identical to the command module of the Soyuz.

"My experience as a fighter pilot stood me in good stead. I felt exactly as if I were in the cockpit of a jet fighter and landed without difficulty. The manual control system of the spacecraft is safe and efficient and can certainly be used on subsequent flights. Our capsule landed easily, thanks to the soft-landing system used earlier with Voskhod-1. It functioned perfectly and has fully proved its merits."

For five hours after Voskhod-2's landing had been recorded in the West. Tass maintained complete silence. The reason became apparent later. The Voskhod-2 had missed its target area and had landed on the snow-covered slopes of the Urals, in the neighbourhood of Perm. The capsule had come down in a dense forest, nearly twenty kilometres from the nearest fire-break. Once the landing site had been located, food and warm clothing for the cosmonauts were dropped. Belyayev and Leonov had to spend the rest of that day and the night that followed in the snow. Next morning a helicopter landed in an open space, and from here rescuers on skis began their search for the two cosmonauts. After they had been found it took another day before they could be brought to safety. It was decided that it would be impossible to pick them up from their landing site. A helicopter again dropped food, and it was not until the next morning, March 21, that cosmonauts and rescuers began the long ski-trek to the firebreak, from where they were finally lifted off by helicopter.

The press conference on the Voskhod-2 flight was held on March 26. On that occasion Professor Mstislav Keldysh, the president of the Academy of Sciences, said that the Voskhod-2 flight opened up wide perspectives: "building space stations, docking spaceships, carrying out astronomical observations in space. It will be possible within the near future to establish space laboratories in orbit round the earth, where specialists in various fields will be able to work. The results of the Voskhod-2 flight form an important step towards the moon and the other heavenly bodies."

In the course of this same press conference Pavel Belyayev* was asked the rather amateur question: "What problems are still to be solved before it will be possible to construct large space stations?"

*Belyayev died on January 10, 1970, following an operation for stomach ulcers.

He replied: "There are many problems, but the chief one is that of bringing several spaceships together in space."

It was therefore not surprising that the objective of the next Russian spaceflight was to find a solution to this very important question.

Voskhod-2 flight summary:

Launched March 18, 1965, 9.00 hrs. Moscow time. Weight 5682 kg. Transmitter frequencies: 18.035, 17.365, 19.996 and 143.625 megahertz. Angle of orbit with equator: 65 degrees. Duration of orbit: 90·94 minutes. Perigee 173; apogee 497·7 km. Duration of flight 26 hrs. 2 mins. 17 seconds. Number of orbits: 18. Landed on March 19, 1965, 11 hrs. 02 mins. 17 seconds.

KOMAROV DIES IN HARNESS

Soyuz-1: new spacecraft – New possibilities – Difficulties in flight – Faulty parachute – Improvements – "A splendid spaceship"

Sunday, April 24, 1967. It is 6 a.m. Moscow time.

A new Russian spaceship has been orbiting the earth for the past 24 hours. The Russians are delighted. For over two years nothing has been heard from the Soviet Union in the field of manned spaceflights. Two years have gone by since the triumphant flight of Voskhod-2 and Alexei Leonov's impressive walk in space. Two years, during which the Americans have launched one Gemini after another with the regularity of clockwork, and have scored one success after another. Will the USSR at last take revenge in this sporting competition, asks the West? It certainly looks like it. Rumours have it that the new Soyuz* is of impressive size. In command is the top Russian cosmonaut: the 40-year-old Vladimir Komarov, the man who had piloted Voskhod-1.

But all does not seem to be well. Tass has been remarkably reticent about the flight. At twelve minutes past six there is at last some more news: "According to a report received from pilot-cosmonaut Komarov at 2.50 he is in good condition. The spacecraft's systems are functioning normally." At that moment only a few people know that Komarov has been having a hard struggle with the recalcitrant Soyuz for the last three orbits. After the eighteenth orbit he at last manages to get the craft under control and to fire the retro-rocket. At a speed of nearly 27000 km per hour the solitary cosmonaut races towards the atmosphere, lying on his back in the capsule. The enormous heat on re-entry presents no difficulty to the new spaceship. Its speed decreases. At a height of seven kilometres, the circular hatch near the top of the cabin is to spring open and the huge, scarlet-edged parachute ejected. The parachute comes out, but it does not deploy. Its cords are tangled. Komarov realises what is happening. He has no ejector-seat

*Soyuz (pronounced "Sayuz") means "Union".

Fig. 14. Soyuz spaceship.
(Key on p. 150).

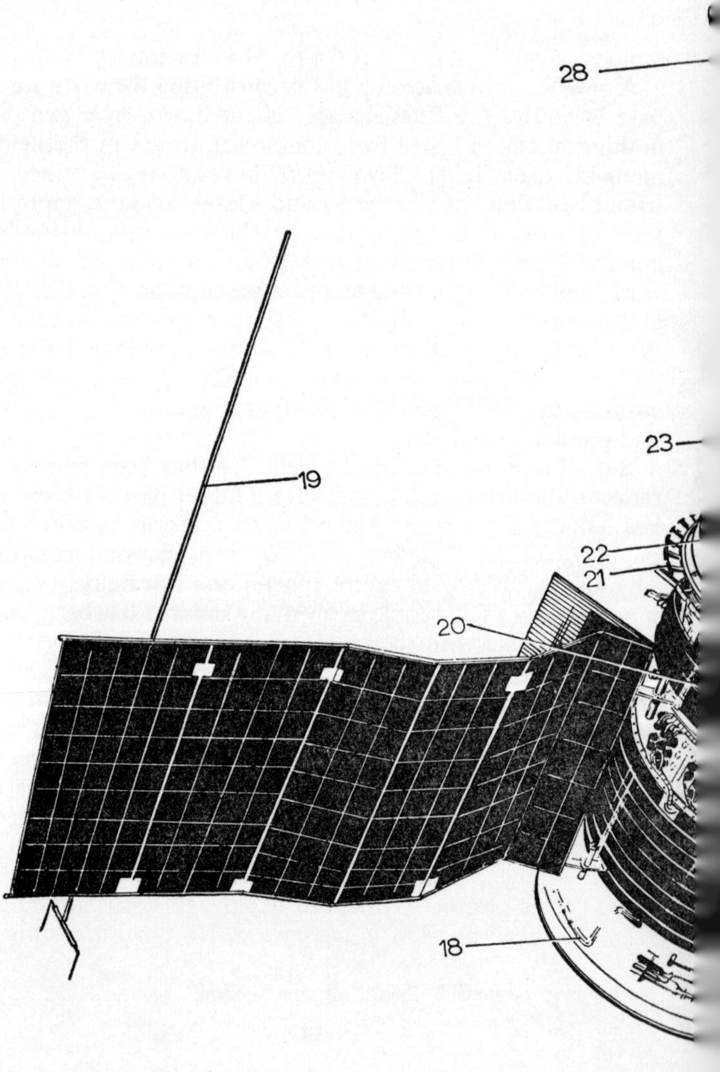

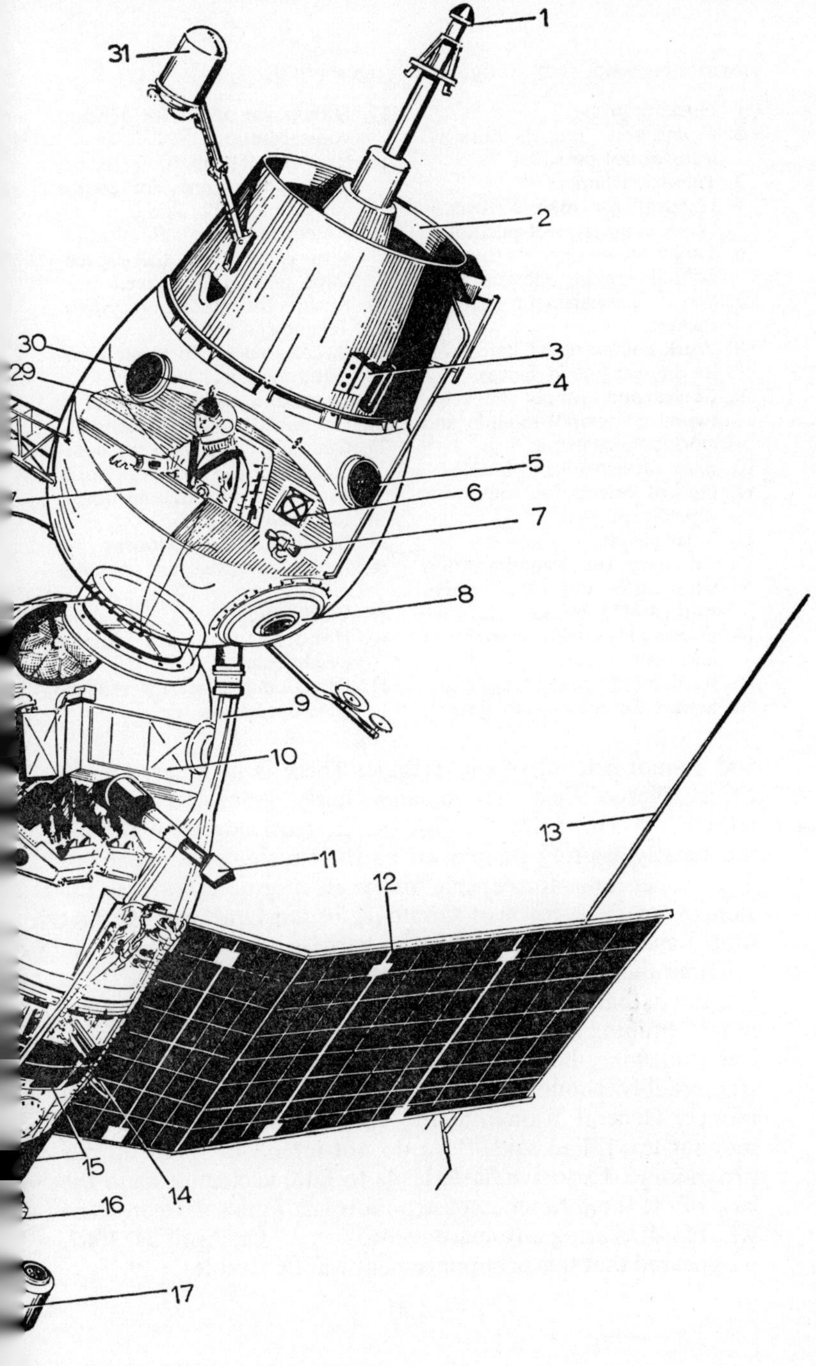
1
31
2
30
29
3
4
5
6
7
8
9
10
13
11
12
15
14
16
17

Soyuz spaceship—key to figure on pages 148-9.

1. Docking probe.
2. Connection tunnel (internal transfer not possible).
3. Television camera.
4. Handrail for space-walkers.
5. Work compartment porthole.
6. Target for rendezvous (homing point in docking operations).
7. Work compartment steering rockets.
8. Work compartment hatch (for leaving capsule in space).
9. Cables and supply lines between command module and work compartment.
10. Back of instrument panel.
11. Optical orientation apparatus ("periscope").
12. Solar panels.
13. Antennae for communication with earth and other spaceships, and for telemetry.
14. Steering rockets for orientation and stabilisation.
15. Radiator temperature control.
16. Sensor for orientation system.
17. Directional aerial for rendezvous operations.
18. Rear end of propulsion section, containing engines for course correction and braking.
19. Antennae for communication with earth and other spaceships, and for telemetry.
20. Sensors for orientation system.
21. Telemetry aerial.
22. Steering rocket to rotate command module.
23. Command module porthole.
24. Command module parachute.
25. Coupling collar between command module and work compartment, with internal hatch.
26. Rendezvous light.
27. Bed in work compartment.
28. Radar antenna for coupling system.
29. Workbench.
30. Handrail (for use during weightlessness).
31. Directional aerial for rendezvous operations.

and cannot get out of the capsule. There is no reserve parachute. Nevertheless he remains fairly calm. "Parachute, para. . . . " These are the last words received from him by the tensely waiting people on earth. Crashing at a speed of 450 km per hour, the capsule makes an enormous crater in the steppes near the town of Orenburg in the Ural. For the first time a space traveller has died in harness. . . .

Three months before, three Americans had lost their lives in a fire in that other new spaceship, the Apollo. This happened during training at ground level. Space travel, too, takes its toll. Each machine, however thoroughly tested, can have defects. Remarkably enough, at the beginning of that tragic April month, General Kamanin, the officer commanding the cosmonaut team, had said: "We do not intend to speed up our programme. Excessive haste leads to fatal accidents, as in the case of the three American astronauts last January. In any case we are not chasing any space records. . . . " On April 24, 1967, it appeared that this pronouncement was debatable. . . .

Operational spaceship

Nevertheless, like Apollo, Soyuz was a promising spaceship, a machine with entirely new potentials. It was the first completely manoeuvrable Russian spacecraft, which would provide the opportunity for rendezvous and docking operations and a craft of totally new conception, even though based on the enormous experience which Sergei Korolyov and his team had gained with the Vostoks and Voskhods.

What does a Soyuz look like? The spaceship consists of three main sections: a work compartment, a command module and a propulsion section. The two former compartments are separated by a hatch which can be hermetically sealed. The work compartment, which can also be used as an air-lock chamber for extra-vehicular activities, has the shape of an extended sphere. Externally it is 2·4 metres in diameter. This section can be depressurised. Behind it is the command module, which on launching provides space for at most three cosmonauts, who can return to earth in it as well. This section contains the guidance control equipment, which enables the crew to manoeuvre the spacecraft and to rendezvous with another spaceship or with a space station. For this purpose the Soyuz can be equipped with a coupling tunnel and a coupling probe.

The command module and work compartment together have a capacity of nine cubic metres, which is not all that much compared with the American Apollo (command module and lunar module together), which provides a living space of fourteen cubic metres. On the other hand, the Soyuz is not intended for travel to the moon. Behind the Soyuz command module comes the cylindrical propulsion unit, to which two large panels with solar cells are attached. These cells convert sunlight into electrical power for the apparatus on board. The propulsion section contains fuel tanks, two rocket engines, each with a thrust of 400 kg—used for course correction and braking operations—and small stabiliser engines.

The shape of the command module, which is approximately that of a cone, is a new element in Russian space travel; the Vostok and Voskhod capsules had the classical spherical shape. Korolyov had several good reasons to get away from the sphere. To begin with, a cosmonaut cannot control the course of a spherical capsule on re-entry through the atmosphere. A sphere presents the same resistance in all directions,

Fig. 15. Service and instrument panels in Soyuz command module.

1. Command signal panel (to left and right of main panel).
2. Off switches for various systems.
3. On switches for various systems.
4. Indicator lights for various systems.
5. Service control switches (communications, landing system, airlock, orientation system, etc.).
6. Main control panel (*Pul't kosmonavtov*).
7. Globe which follows the spacecraft's motions in orbit, constantly indicating to cosmonauts their position over the earth's surface—with accessories.
8. Globe adjuster knob.
9. Orbit counter.
10. Orbiting time indicator.
11. Orbit counter adjuster.
12. Indicator lights for various systems.
13. T.V. brightness adjuster.
14. Television screen (for use, e.g., during rendezvous operations).
15. Television contrast adjuster knob.
16. Flight programme projection screen.
17. Digital information panel.
18. Pressure meter for command module, work space and spacesuits.
19. Service knobs (see 5).

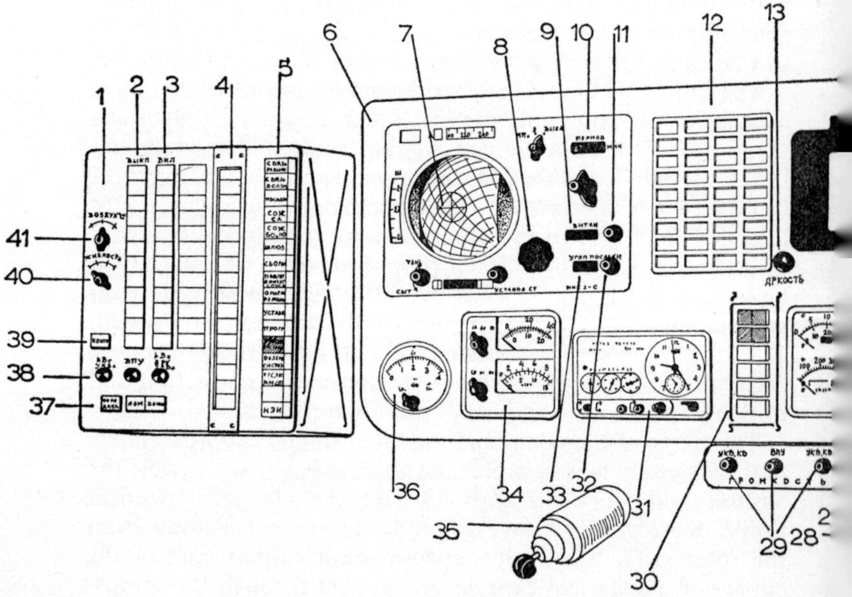

20. Signal lights (see 4).
21. On switches (see 3).
22. Off switches (see 2).
23. Control knob for signal lights (see 39).
24. Adjuster knob for communications apparatus (see 38).
25. Microphone and headphone switches (see 37).
26. Orientation device, used chiefly in manual landings (see "Vzor" in Vostok).
27. Speedometer and Odometer.
28. Manual stabiliser (yaw, pitch and roll).
29. Radio adjuster knobs.
30. Switches for signal lights.
31. Chronometers.
32. Adjuster knob for 23.
33. Angle of descent indicator.
34. Temperature and air-pressure meters for all three compartments (command module, work space and propulsion unit).
35. Manual control (for course correction).
36. Current- and Voltage meters.
37. Microphone and headphone switches (see 25).
38. Adjuster knobs for communication apparatus (see 24).
39. Control knob for signal lights (see 23).
40. Cooling system temperature control.
41. Cabin temperature control.

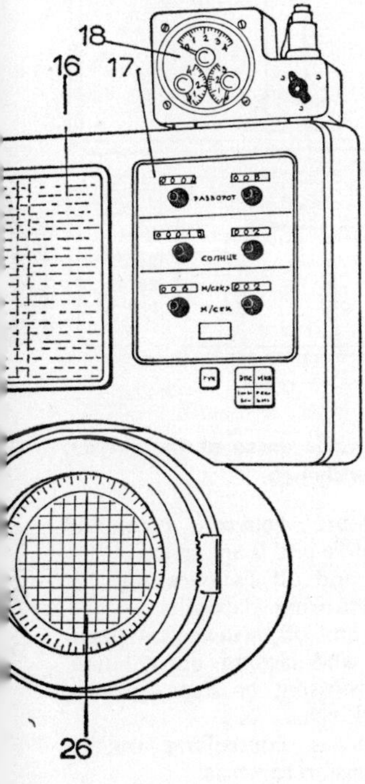

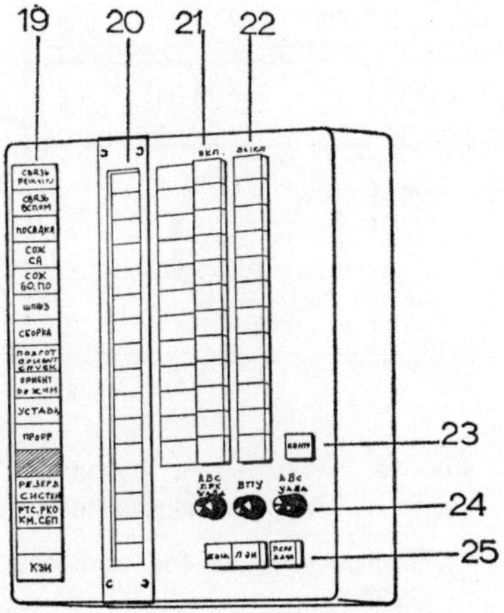

and can therefore only follow a "ballistic trajectory". Course deviations which may occur during the return to earth, cannot be corrected, so that the predetermined landing area can only be reached by approximation. In the case of a cone-shaped capsule, with a practically flat base serving as a heat-shield, the line of descent is to a large degree dependent on the angle between heat-shield and flight path. By means of small, inbuilt jets at the base of the spaceship, the cosmonauts can vary this angle, thus increasing or decreasing the lift. In this way they can to some extent determine their point of landing, although of course the position at which the retro-rocket was fired and the magnitude of the braking force are decisive.

During return in a cone-shaped capsule it is, moreover, easier to control the rate of deceleration and consequent stress,

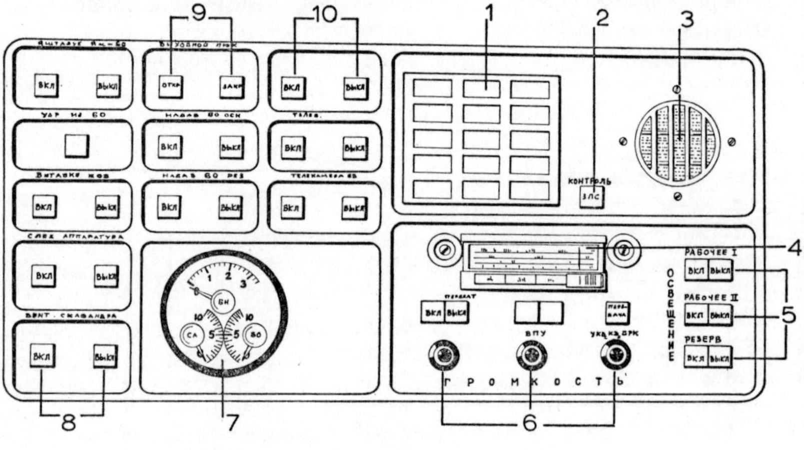

Fig. 16. Service- and instrument panel in work space of the Soyuz. The panel is situated above the workbench.

1. Indicator lights for on-board systems.
2. Indicator light switch.
3. Loudspeaker.
4. FM, medium wave and long wave radio.
5. On and off light switches.
6. Radio knobs.
7. Meter which records, among other things, differences in pressure between command module and work space.
8. On and off switches, e.g. for pressurising spacesuits.
9. On and off switches for opening and closing outer hatch, pressurising or depressurising work space.
10. Switches controlling outside television cameras.

which is, of course, of particular importance when returning at great speed, for instance from a flight to the moon. No traveller to the moon could survive return in a spherical capsule.

Compared to its predecessors, the Vostok and the Voskhod, the Soyuz is the first truly operational Russian spaceship: a manoeuvrable craft, capable of sustaining a three-man crew for a whole month. The work compartment can serve as air-lock chamber and for experiments, and the command module can transport "personnel". Its possible applications are numerous, even though its manoeuvring capacity is still somewhat limited. The maximum height it can reach is 1300 km. The two 400 kg engines used for course correction serve also as retro-rockets. The thrust of one of these engines suffices to bring the Soyuz back to earth; the other serves as reserve.

I am calm

Saturday, April 23. It is still dark at Baikonur, where late snow covers the platforms surrounding the modified launch pad, now adapted for sending up the RNS (*Rakyeta Nosityel Soyuz*), more powerful than its predecessors.* The rocket is bathed in the light of floodlights mounted on tall posts. It is still early. In Moscow it is one o'clock at night; local time is 6 a.m.

The platform is buzzing with voices. The pale blue cosmonaut bus appears from the darkness of the steppe. The technicians, the members of the State Commission, several cosmonauts and a number of journalists who, a moment ago, were interested only in the rocket, approach the bus, which has stopped in front of the launching pad. Vladimir Komarov gets out. He is wearing the same plain blue and grey training suit as on the Voskhod-1 flight. The only part of his outfit which betrays his profession is a small white helmet with headphones and microphones.

Vladimir Komarov is the first Russian to go into space for the second time. It is said that there is no better test pilot. Shortly after his flight in Voskhod-1 he was medically rejected for further participation in the space programme, but thanks

*See: "The Super-Rocket Myth", p. 54.

to his perseverance his physical condition improved and he was allowed to continue training. . . .

Now he is ready for his second voyage in space. "I am calm. I'm perfectly O.K.," Vladimir tells the journalists, smiling. Then he steps into the lift and is swept upwards to the waiting Soyuz-1. It is two hours before lift-off: 1.35 Moscow time.

03.05: "*Ja Rubin, ja Rubin.*" (This is Rubin, this is Rubin), Komarov's voice comes from the loudspeakers in the block-house. "Let's check the time." The clock on the instrument panel in the command capsule is corrected. Now the solitary cosmonaut has concluded the numerous technical checks.

03.35: The moment of lift-off has come. "*Zazhiganiye*," says the officer in charge of the launching, who is watching through a periscope, and enormous flames appear from the base of the rocket. The tentacles of the launching towers are pulled back. Soyuz-1 is on its way. "*Vsyo normalno na bortu*" (all well on board), reports Komarov. A few minutes later the Soyuz is in orbit round the earth at between 201 and 244 km. The angle of its trajectory with the equator is 51 degrees 42 minutes, the same as in the case of the Russian moon probes. Calmly Komarov relays the information provided by the instruments on the panel in front of him. The spaceship is functioning well. The cosmonaut is cheerful.

04.00. Pavel Popovich, who piloted Vostok-4, telephones Valya Komarov in "Star Village" to tell her about the launching. She did not know that Vladimir would go up in Soyuz-1. "My husband never tells me when he goes on a business trip," she tells a reporter a moment later.

She is obviously nervous, much more so than three years ago when her husband was piloting the first Voskhod. She calms down a little after she has heard his voice over the radio. "Volodya spoke very calmly, just as if he were at home," she says. "Everything appears to be going well. I hope he can complete the programme and that he will return safely. We are waiting for him in suspense." ("We" are Valya and her two children: nine-year-old daughter Ira and fifteen-year-old son Yevgeni.)

At this time Komarov is subjecting the Soyuz-1, racing round the earth at a speed of 28 000 km per hour, to extensive tests. Both the manual and the automatic steering gear are functioning well. Obediently the colossus, seven metres long and weighing six tons, revolves on its axis. The guidance control engines work well, and so do the two rocket engines in the propulsion unit. Vladimir switches on the engines a few times, putting his Soyuz into a higher orbit. The spacecraft proves easy to handle. There is not the slightest cause for alarm. Vladimir does not confine himself to technical matters; he congratulates the Soviet people with the approaching fiftieth anniversary of the Revolution and generously scatters messages for the countries over which he is flying.

10.00 hrs. The Soyuz-1 has nearly completed its fifth orbit. The flight is still going according to plan. Komarov reports that the temperature in the cabin is 16 degrees Celsius, the pressure of the artificial atmosphere is exactly 76 cm.

13.30 hrs. The spaceship will be out of radio contact with the tracking stations in the Soviet Union until 21.20: the earth revolves under the spacecraft's trajectory. Komarov will dictate his experiences into a small tape-recorder.

21.20. Radio contact has been re-established. Komarov reports. For the past eight hours he has been sleeping well. There are still no problems.

23.00 hrs. Komarov completes his thirteenth orbit. All is still going well.

April 24

02.50 hrs. Komarov reports that he is feeling fine. The spaceship functions as expected. Tass publishes this report at 06.12.

About 03.00 hrs. A first hint of problems. There are difficulties with the stabilisation of the spaceship, which is making uncontrolled movements round its centre of gravity. Communication with earth is also not faultless, probably because of the fact that the antennae are constantly pointing in the wrong directions. Komarov also reports that direction control is using up far more fuel than was anticipated.

"The flight control centre at Baikonur reported difficulties in communication during the fifteenth and sixteenth orbit," Radio Moscow was to announce later. "Certain difficulties arose in connection with position control in relation to the earth."

Positioning is of course of the greatest importance in connection with the spacecraft's return. At the time when the retro-rocket is fired, the position in relation to the earth must be exactly right. In an extreme case the spaceship's speed could be increased rather than decreased, with the result that it would get into a higher orbit. However, as with Voskhod-2, the automatic orientation system is not functioning properly. With Mission Control's permission Komarov decides to make a manual landing after the seventeenth orbit. But when the critical moment arrives, he does not succeed in putting the recalcitrant spacecraft into the correct position. One more orbit. At the end of the eighteenth orbit he succeeds. The cone-shaped capsule, now separated from the propulsion unit and the work compartment, races towards the atmosphere.

06.00 hrs. The capsule withstands overheating perfectly. It seems as if the worst is over. But the capsule is still revolving on its axis. At a height of seven kilometres the parachute should come out. The hatch flies off, but as a result of the craft's revolutions, the cords get entangled. The parachute barely unfolds. The capsule, although already considerably slowed down by atmospheric friction, does not decelerate further. Komarov remains in contact with the earth as long as possible. "His final reports were a model of clear, concise information, of self-control and calm," the journalist Sergei Borzyenko wrote in Pravda the next day. "He struggled to salvage the spaceship entrusted to him up to the very end."

06.15 hrs. With a speed of 450 km per hour the three-ton capsule crashes into the hills near the town of Orenburg in the Urals. The machine bursts into flames. Rescue- and recovery-teams are soon on the spot, but little can be done. The capsule's strong hull is largely intact, but the interior is completely destroyed. . . .

A long period of silence follows. Not until 3.17 that afternoon does Tass publish the following official communiqué:

"As previously announced, the new spaceship Soyuz-1 was launched in the Soviet Union on April 23, 1967 for a test flight in an orbit round the earth; the spaceship was piloted by pilot-cosmonaut of the USSR, Hero of the Soviet Union, Colonel of the Engineers, Komarov, Vladimir Michailovich.

"In the course of the test flight, which lasted more than a day, Komarov completed the flight programme, testing the system of the new spaceship and in addition carrying out the scientific experiments which had been planned. During the flight, Komarov manoeuvred the spaceship, tested its main systems under varying conditions and gave an expert opinion on the technical qualities of the new spacecraft. When on April 24 the test programme had been completed, it was suggested to him that he should conclude the flight and land. After having carried out all the operations connected with the preparations for the descent, the spaceship safely accomplished the most difficult and important part of the deceleration in the atmosphere and the first cosmic velocity was successfully superseded. However, when the main parachute was deployed at a height of seven kilometres, the spaceship, according to preliminary reports, crashed at great speed as a result of the parachute cords getting entangled, killing Komarov.

"The untimely death of this outstanding cosmonaut, engineer-pilot of spaceships Vladimir Michailovich Komarov, constitutes a great loss for the Soviet people. By his activities in the field of spacecraft testing Vladimir Michailovich Komarov contributed significantly to the development and execution of space technology."

A splendid spaceship

Some people in the West doubted whether Tass' version of the accident was the true one. The suspicion was voiced that Komarov had died in space of a heart attack, that the capsule had dived into its own parachute, etc., etc. All this was sheer conjecture. As had happened before, Tass did not give all the facts, but the details which were announced were correct. NASA, the American Space Administration, confirmed that there was no reason to doubt the Tass report. Possibly the Americans possessed more information than was apparent.

A few days after the fatal accident the urn containing Komarov's ashes was immured in the Kremlin wall, an honour granted only to the very greatest Soviet citizens. The American government had wanted to send a deputation of astronauts, but the Russians informed them that the ceremony would be of a strictly national character.

Messages of sympathy were received from all over the world. The Russians found out for the first time that a space drama is for the West by no means an occasion for what is called "counter-propaganda". The fact that they had obviously feared this was shown by what a Russian said to me some months later: "We were very relieved. We received nothing but genuinely human reactions and messages of sympathy."

The most significant commentary was perhaps that of James Webb, who until 1969 was Administrator of NASA. He urged that there should be closer co-operation. "This might have prevented the death of Komarov and of the three American Apollo-astronauts," he said.

Immediately after the disaster a State Commission was appointed to investigate the exact cause of the accident—just as had happened a few months earlier in the United States. The Soviet commission had no easy task. True, it had the heavily damaged command module of the Soyuz, but the work compartment and the all-important propulsion unit had burnt up on re-entry into the atmosphere, which they were not equipped to survive. To facilitate the investigations, a new Soyuz was taken entirely to pieces.

It was suggested in the West that the Russian manned space programme would be delayed as a result of the disaster, but the facts proved otherwise. The commission succeeded very soon in establishing the true course of events and made recommendations for effective changes which were easily introduced. The amendments chiefly concerned the orientation system and the parachute section. A small stabiliser parachute was added to counteract possible rolling of the capsule before the main parachute deployed. In addition a reserve main parachute was mounted. A complete revision of the entire Soyuz construction was unnecessary. "The Soyuz is a splendid spaceship which will undoubtedly be used for further flights," Yuri Gagarin declared in response to suggestions that the Soyuz had proved unserviceable. As early as July 17, 1967 a new unmanned Soyuz, Kosmos-169, was sent up to test the efficacy of the amendments which had been incorporated.

Summary of Soyuz-1 flight

Launched: April 23, 1967, 03.35 hrs. Moscow time. Initial weight approx. 6 tons. Angle of trajectory with the equator 51·7 degrees. Orbiting time 88·6 mins. Perigee 201 km, apogee 224 km. Duration of flight: 26 hrs. 40 mins. Crash landing on April 24, 1967, 06.15 Moscow time.

Details of Soyuz spaceship

Total length: 7·81 m (inclusive of docking unit). Command module: length 2·00 m, diameter 2·19 m (at base). Work space: 2·27 m, maximum diameter 2·19 m. Propulsion unit: length 2·76 m, diameter 2·80 m (at base).

SOYUZ-2 AND -3: FIRST RENDEZVOUS IN SPACE

Beregovoi: the oldest space traveller – Rendezvous with Soyuz-2 – Sleeping on the sofa – No docking yet – Accurate landing

"No, vitality and health are not dependent on age. Cheerfulness, supple muscles and strength and dexterity are achieved by systematic physical training," wrote *Red Star* on December 21, 1968 under a photograph showing Georgi Beregovoi in training. Beregovoi, aged 47, was at that time the oldest cosmonaut in the Russian team and the oldest space traveller in the world.

Beregovoi already was a Hero of the Soviet Union before his first spaceflight, and had been decorated several times for his wartime service. He it was who, 18 months after the Soyuz-1 disaster, flew another Soyuz and thus inaugurated a new period of successful manned Russian spaceflights. On October 26, 1968, his Soyuz-3 had a rendezvous in space with Soyuz-2, an unmanned spacecraft launched a day earlier, thus laying the foundations for the subsequent extensive rendezvous and docking operations of the Soyuz craft. His flight had been preceded by several unmanned Soyuz flights and by two impressive docking experiments: the automatic coupling of Kosmos-186 and -188 on October 30, 1967, and a similar experiment with Kosmos-212 and -213 on April 15, 1968.

The launching of Soyuz-2 on October 25 was not immediately announced. The satellite was put into orbit at between 185 and 224 km. The angle of its trajectory with the equator was 51·7 degrees as had been the case with the first Soyuz. The same applied to Soyuz-3 with Beregovoi on board. His maximum height was 225 km—only one km more than that of Soyuz-2. His perigee, however, was higher: 205 km, entirely according to plan.

Already, in its first orbit Soyuz-3 started chasing No. 2. The main part of the operation was carried out by the same excellent automatic pilot which had previously brought the two Kosmos satellites together. Not until the two Soyuz craft had approached within 200 metres of each other did Beregovoi take control to guide his "ship" to within a few metres of Soyuz-2. The fact that the cosmonaut had carried out his task well was clearly seen on earth by means of television; Soyuz-2 was perfectly lined up with its twin craft. Beregovoi could have linked the two craft, but this operation had not been planned on this occasion. The two spaceships then moved apart once more. When Soyuz-3 had entered on its fifth orbit, Beregovoi crawled through the hatch into the work compartment of his spacecraft and went to sleep on the sofa. "I felt perfectly at home," he said later and this was borne out by the medical information obtained. Beregovoi slept very peacefully. His pulse remained within the limits of 56 and 60; his breathing frequency was 16. During lift-off both these parameters had been considerably higher, though by no means alarming: 100 and 30 respectively. Practically as soon as he went into orbit Beregovoi had been perfectly calm again.

When he awoke, the cosmonaut spent 25 minutes doing physical exercises and having breakfast.

By 6.50 p.m. Moscow time on October 27, Soyuz-3 had completed 22 orbits and Soyuz-2 no fewer than 38. Again earth instructed the two spacecraft to approach each other. At that moment they were hundreds of kilometres apart. Soyuz-2 was still orbiting at a height between 185 and 224 km, for this spacecraft was, as intended, playing an entirely passive role. As a result of the manoeuvres connected with the first rendezvous operation, the trajectory of the manned capsule was different from the original one. Now the two engines of the Soyuz-3 were fired again to correct its course in order to make it approach Soyuz-2 once more. As before, the rendezvous operation was controlled from earth. The radar antenna, mounted on a large mast linked to the cabin, measured the decreasing distance between Soyuz-3 and -2. The information was immediately fed into the on-board computer in the command module, which in no time at all passed "orders" for positioning and propelling the craft. When the distance between the two spaceships had been reduced to 200 metres, Beregovoi once more took control. On the television screen facing him he could see the distance between the two capsules. At that moment he could no longer see his objective through the portholes to right and left. And in front of his cabin was the work compartment, so the periscope in his cabin was no superfluous luxury.

Beregovoi would have been able to carry out the entire operation by manual control. He need only have operated one switch. As before, the radar antenna would have fed information into the computer, which would then, instead of activating the engines itself, have shown the details required for course correction on the instrument panel in front of the pilot, indicating the correct course for him to fly.

The Russians have never explained in so many words why, on this occasion, no docking operation was attempted. Naturally some people suggested that there had been a technical hitch, but this is unlikely. The Russians had already twice successfully linked unmanned Soyuz craft, thus proving that they had mastered the techniques of cosmic manoeuvres and docking. Linking two spacecraft could therefore hardly present a problem. On the other hand, the Soyuz-1 disaster was still fresh in the memory of Russians and they did not want to let the solitary pilot on board Soyuz-3 incur unnecessary risks. Beregovoi's main task was to test the Soyuz-3, both as regards its on-board equipment and its ease of handling. It was the first time that the spacecraft was piloted manually—if only for short periods. This in itself was enough. A link-up of manned spaceships could wait till the next flight, when more than one cosmonaut would be sent up.*

Soon after the second rendezvous, Beregovoi gave another television performance in his Soyuz. Millions of people on earth could see that he was enjoying his high-altitude trip. The Soyuz appeared to be a comfortable spacecraft and with its curtained round portholes almost gave an impression of domesticity. Beregovoi was wearing the well-known woollen training suit and a simple white helmet containing earphones and microphones. In the West viewers waited expectantly for possible spectacular experiments during the rest of the flight, but these were not forthcoming. On October 28, at 10.25 Moscow time, a retro-rocket of the unmanned Soyuz-2 was fired and one partner left the field. At 10.51 the cone-shaped capsule dived into the dense layers of the atmosphere and landed completely intact at the predetermined place. In

*In this way the Russians could also be the first to link up *manned* spaceships. The American Gemini astronauts had already demonstrated a link-up with an unmanned spacecraft.

Moscow it was announced that the Soyuz-2 had carried out its task and that its programme had been entirely completed.

Beregovoi, who meanwhile cheerfully continued to circle the earth in his Soyuz-3, now had more time to devote himself to scientific experiments and observations. During his 33rd orbit he reported three forest fires and near the equator he found a heavy thunderstorm. He also saw several hurricanes as well as whirlwinds with their typical funnel-shaped whirlpools in the cloud layer. These observations proved once again that a more or less permanent observation post at high altitude could be of immense importance. He also observed the stars.

Soon after midday on October 28, the solitary cosmonaut gave another television show. He told the viewers something about the equipment and the instruments on board the spaceship and gave a demonstration of weightlessness by letting a television camera float freely. During his 33rd orbit Beregovoi also switched on the engines, once again changing the Soyuz' course. After this manoeuvre the apogee was 244 km and the perigee 199 km. After this he put his capsule in the correct position once more, so that the solar panels, which resembled wings, were directed towards the sun.

Beregovoi's fourth working day started at 3.45 Moscow time on October 29. "I've slept well," he announced via the radio. As behoves a good cosmonaut, he first did his morning exercises and then enjoyed his breakfast. A brief check of the spaceship showed that everything was functioning normally. Cabin pressure was still 770 mm mercury, the temperature was a comfortable 18 degrees Celsius. The doctors were pleased with the cosmonaut's exceptionally good appetite. He quickly ate his lunch, consisting of fish, chicken, biscuits, milk chocolate and prunes. Telemetric observations showed that during this fourth day his pulse varied between 63 and 68 and his breathing frequency between 12 and 15.

That afternoon, at 12.15 Moscow time, Beregovoi again gave a live television show. This time he not only showed the interior of his Soyuz—he also gave the viewers a sight of the earth through one of the portholes.

At 10.25 on Wednesday, October 30, Beregovoi landed in his capsule near Karaganda (Kazakhstan). Like Soyuz-2, after the retro-rocket was fired (for 145 seconds), the capsule had re-entered the atmosphere and, using "lift" had flown on a trajectory which subjected the cosmonaut to no more than two or three times his own weight. On a purely ballistic re-entry, such as that of the Vostoks, this was eight or nine times. The flight had lasted exactly 3 days, 22 hours and 51 minutes. The Soyuz, which had made such a tragic début with Komarov, had proved to be a good spacecraft.

"Georgi Beregovoi has carried out an immense research programme," Professor Mstislav Keldysh said at a press conference held on November 5. "Apart from his geophysical, astronomical and bio-medical observations, the cosmonaut noted typhoons, whirlwinds and forest fires. We are already using unmanned satellites for radio communications, weather forecasts, navigation and other purposes. Manned flights offer even greater opportunities to make practical use of space."

The cosmonaut Feokistov, the engineer who had flown in Voskhod-1, supplied several important details about the construction of the Soyuz, which have already been mentioned above. He drew particular attention to the fact that the new command module could more or less determine its own course, which made it easier to withstand excessive stress and enabled the cosmonaut to remain active. Moreover, the landing site could be approached more accurately. "I could see the face of a member of the recovery team through my porthole before I had time to get out of the capsule," Beregovoi related enthusiastically.

A stabiliser parachute had opened at nine km, he said, and at seven km the main parachute had come out. A reserve parachute was carried in case the main chute did not function. At a height of exactly one metre the solid fuel rockets burnt, so that the spaceship landed exceptionally softly. Beregovoi implied that the entire landing operation had been carried out by the automatic pilot.

Professor Mstislav Keldysh was asked a number of questions. By way of introduction he said: "The Soyuz craft are equipped for solving many problems concerning space exploration and they will of course make further flights with various scientific objectives. Naturally these spaceships are constructed *for the completion of a specific cosmic programme*, not for one single flight." This was the first indication that the Soyuz was to play a part in a much more extensive project.

"These vehicles are equipped with automatic docking devices, which have already been tested in satellites of the

Kosmos type," said Keldysh. "These devices will be used on subsequent flights as well."

An East German television commentator asked: "Is there a difference between the space programmes of America and the Soviet Union?"

Keldysh: "Of course there are differences. I might say that at times these programmes complement each other in an interesting manner. Take for instance the flight to Venus. We sent a spacecraft which descended gradually in the atmosphere and took measurements. The Americans launched their Mariner, which flew close to the planet and also made measurements. The details thus obtained complemented each other."

A correspondent of United Press International posed the inevitable question: "Can the Soyuz be used for a flight to the moon?"

Keldysh: "This spacecraft has been designed for flights in low earth-orbit."

Another question was: "What are the limitations of automatic vehicles in space exploration? What part do manned spacecraft have to play?"

Keldysh: "Generally speaking, the possible applications of automatic space vehicles are still expanding. They can be used for an increasing variety of missions. But man will always want to take part in scientific research, the more so since robots can never entirely replace people. At the same time the *current** Soviet space programme lays great emphasis on the use of automatic spacecraft for exploring the moon and the planets. These result in photographs of the panorama and the surface of the moon. We have learned that the moon has no magnetic field. Unmanned spacecraft have also taught us a great deal about the atmosphere of Venus. Moreover, practical applications close to earth, such as those of radio communication, television and weather forecasts are realised to a certain extent by robots."

Keldysh further stated among other things: "There is speculation as to whether there is competition in space between the United States and the Soviet Union. My reply to this is as follows: If two scientists are working in the same field, can there be any question of competition? To some extent this is probably the case. But this must never be allowed to become the predominant factor, for if everything is done for the sake of

*Author's italics.

winning the competition or the race, you may as well forget about science."

Summary

Soyuz-2. *Launched on October 25, 1968, 10.30 Moscow time. Initial weight approx. 6 tons. Angle of trajectory with equator: 51·7 degrees. Orbiting time: 88·5 mins. Perigee 185 km, apogee 224 km. Total duration of flight: 72 hours 30 mins. Landed on October 28, 1968 at 11 o'clock Moscow time.*

Soyuz-3. *Launched on October 26, 1968, at 11.34 Moscow time. Initial weight approx. 6 tons. Angle of trajectory with equator: 51·7 degrees. Orbiting time: 88·6 mins. Perigee 205 km, apogee 225 km. Total duration of flight: 94 hours, 51 mins. Landed on October 30, 1968 at 10.25 Moscow time.*

25. (*Above*) *Left* Luna-16 on the moon. The core drilling mechanism can be seen to the right of the spherical top section. *Right* The ascent stage, carrying lunar samples, leaves the moon.

(*Below*) Luna-16 after landing. The flexible cylinders serve to righten the capsule if it lands in the wrong position. The antennae of the homing device are also visible.

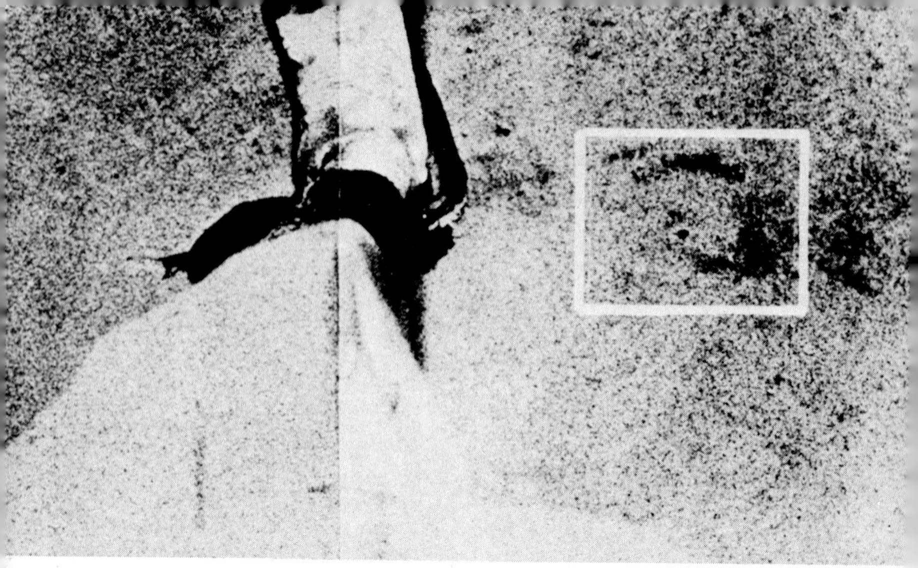

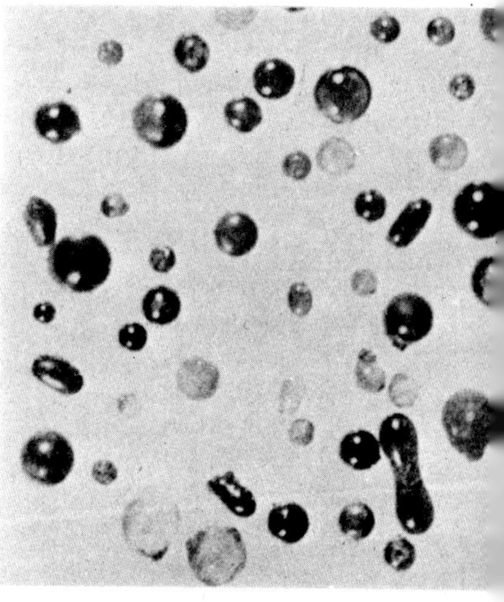

26. (*Above*) One of Luna-20's telephoto cameras took this picture after a lunar sample had been gathered. *Inset* the spot where the sample was taken.
(*Centre*) *Left* Loose rocks from the Luna-16 sample. *Right* Variously shaped and sized glass globules from Luna-16 sample.

(*Below*) Overall view of the sample of lunar material brought back by Luna-16. At lower levels (working from left to right) up to a depth of 35 cm the sample contains more coarse-grained particles.

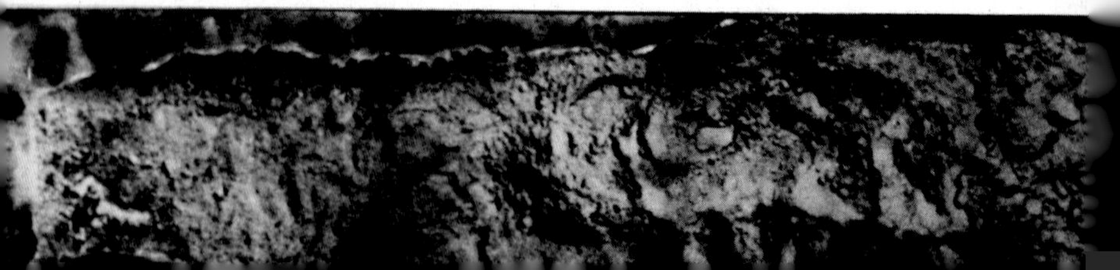

27. (*Above*) *Left* Lunokhod-1 runs down the Luna-17 gangway on to the moon's surface. The photograph was taken by the Lunokhod. One of its wheels can be seen. *Right* The vehicle's wheels leave a clear track on the surface.

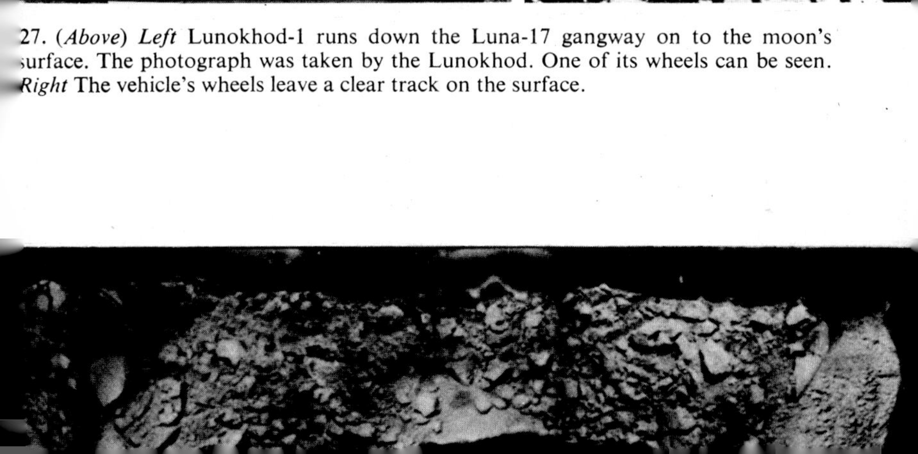

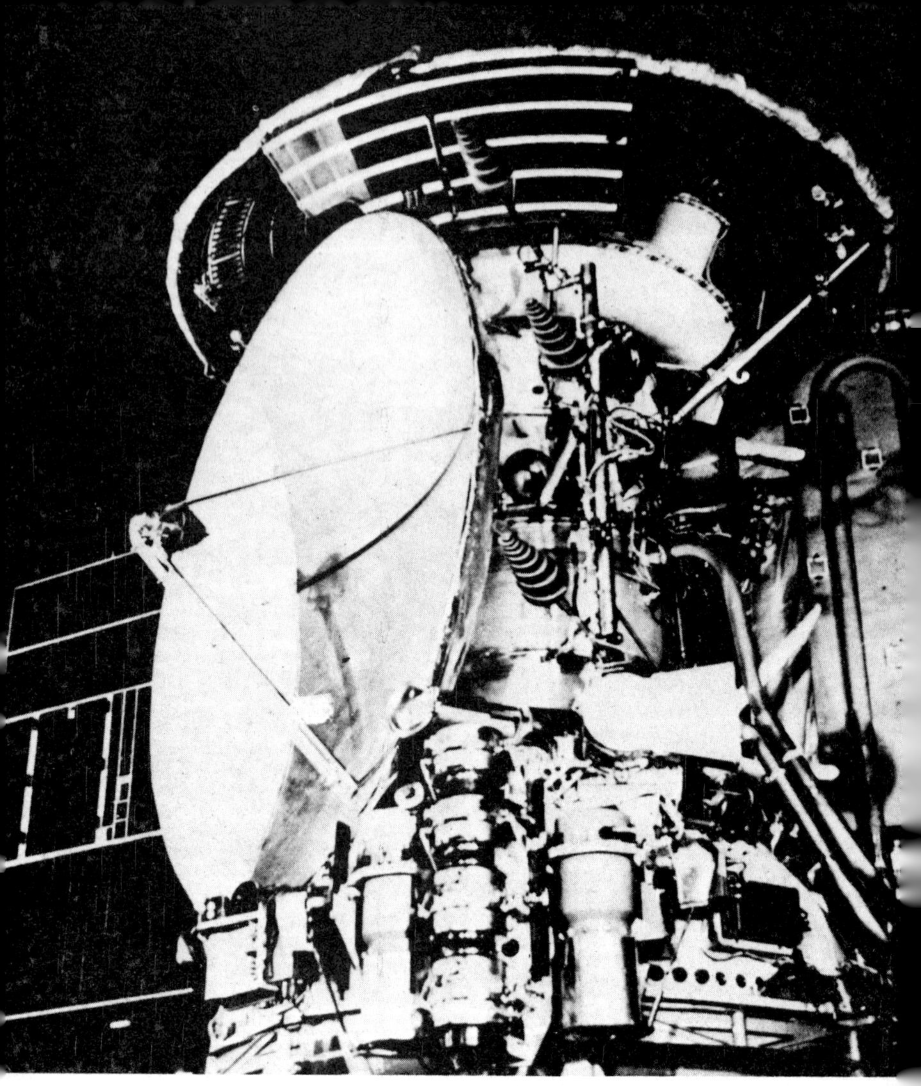

28. (*Above*) Mars-3 before launching. On the left one of the two solar panels, deployed. The parabolic antenna for communication with the earth is of enormous size. At the top the conical heatshield together with the capsule, concealed beneath, which made a soft landing on Mars.

(*Below*) The Lunokhod-1 repeatedly traversed moon craters. The crater shown here has a diameter of 5 metres. To the left is a camera used for registering vertical measurements.

HIGH LEVEL TRANSFER

Experimental space station – A wintry start – Soyuz-4 and -5 link up – Two cosmonauts change capsule – "A first-class ship" – First step towards large space stations

On January 16, 1969, only a few months before man first set foot on the moon, the Soviet Union took a considerable step towards the realisation of a manned space station in orbit round the earth—the main objective of Moscow's space programme. On that day, 11.20 Moscow time, a "first experimental space station", as Tass called it, was established and thus the rumours which for weeks had been gaining ground in the Russian capital were confirmed. But the most spectacular event had yet to come: while Soyuz-4 and -5, linked together, orbited the earth, cosmonauts Yevgeni Khrunov and Alexei Yeliseyev stepped out of Soyuz-5, leaving their colleague and commander Boris Volinov behind. For 37 minutes they carried out a variety of experiments and then transferred to Soyuz-4, where Vladimir Shatalov was waiting for them. This was the first time a change-over had been effected in space. After four hours and 35 minutes the two capsules were disconnected. Soyuz-4 landed at 9.53 the next morning and Soyuz-5 followed a day later.

"This experiment," Tass concluded rightly, "makes it possible to carry out operations in space such as replacing crew members of permanent space stations and rescuing cosmonauts in emergencies."

Vladimir Shatalov, who was entrusted with the "overall command" of the spectacular experiment was, after Komarov, the second Russian cosmonaut to make two spaceflights within a short space of time: he was to take part in the "troika flight" of Soyuz-6, -7 and -8 as well. General Kamanin, who was in charge of the team of cosmonauts, had his reasons: like Komarov, Shatalov was technically very well grounded; he was a "natural" pilot and exceptionally well equipped to carry out experiments calmly which were not without danger. Until the start of Soyuz-3, the 41-year-old Shatalov had been the stand-in for that other exceptional test pilot, Georgi Beregovoi.

A wintry start

It was freezing hard when at 10.39 Moscow time on January 14, Soyuz-4 lifted off from the launching pad. The launching site was covered with a thick layer of snow. Up till this time, "experts" in the west had thought that the Soviets could launch manned spacecraft only at other seasons; this theory proved to be unfounded. A few days before the departure of Soyuz-4 the Russians had launched their Venera-6 . . . spaceflight activities out there in Kazakhstan were certainly not governed by the weather.

"Zarya,* this is Amur. Everything is in order, everything perfect." Those were Shatalov's first words on his voyage into space. Soon after arriving in orbit at a height of between 173 and 225 km he said simply: "I like it up here." He then started to check the systems of his 6-ton Soyuz-4, which was running like clockwork, and to make a start on the preparations for the rendezvous which was to take place the following day.

Shortly before travelling out of radio reach of the ground stations and taking his planned rest period, Shatalov dutifully reported that the first day's programme had been successfully concluded. He woke at 3 a.m., an hour sooner than was planned, when he could not yet talk to Mission Control. As soon as radio contact had been re-established, he reported that he had slept well and would now like to have his *zavtrak* (breakfast).

Transfer

Meanwhile Moscow and the rest of the world were buzzing with rumours. From the fact that Shatalov had corrected his trajectory to a more circular orbit, that is, to between 207 and 237 km, it might be concluded that a possible rendezvous with another spaceship, as yet to be launched, was planned. The course correction had taken place in the fifth orbit, at 16.35 Moscow time. It was further suspected that a transfer might take place, which would fill the two empty seats which had been clearly visible during a television transmission. For once the rumours proved to be more or less correct. At 9.46 that morning the Soyuz-5 pierced the sky over Baikonur to go

*"Zarya" (Dawn) is the call name of Mission Control.

into orbit at between 200 and 230 km. On board were Boris Volinov, Alexei Yeliseyev and Yevgeny Khrunov.

Although immediately after the departure of Soyuz-5 the two spaceships were fairly close together, a large number of manoeuvres had to be carried out to make their orbits coincide. Both capsules underwent a great many course corrections. Soyuz-5, which was to be the passive partner in the docking operation, went into a new orbit with an apogee of 253 km and a perigee of 211 km. By the morning of January 16, Soyuz-4 had arrived in practically the same orbit: between 253 km and 210 km.

With the aid of the rendezvous radar equipment and the onboard computer of the Soyuz-4, Shatalov finally concluded the last part of the cosmic chase. From 10.37 Moscow time onwards the automatic guidance system carried out the main rendezvous operations, until the two capsules were within a hundred metres of each other. It was then that Shatalov took over manual control.

Slowly the two capsules approached each other. Shatalov in Amur reported to his colleague Volinov in Baikal and to earth (Zarya):

Shatalov: "Everything normal. Distance forty metres. Speed almost zero. We continue to approach."

Ground Control: "Understood. We are tracking you."

Volinov: "O.K. . . . Amur . . . Zarya, here is Baikal. I am receiving you perfectly. Distance forty. Direction control excellent. The only thing I don't understand is why my 'docking' control light has gone out, and 'contact' is on. . . ."

Ground Control: "That is all right. That's how it should be."

Shatalov: "Can I start docking?"

Ground Control: "Permission to dock. This is Zarya. Tell us, if possible, briefly what you are doing."

Shatalov: "Understood. I have Baikal on my screen. Speed 0·25, distance 30. Everything normal."

Shatalov: "Distance 25, speed 0·25."

Ground Control: "Everything normal."

Shatalov: "Distance 20, speed 0·25."

Ground Control: "We are tracking you."

Shatalov: "Distance 10, speed 0·25."

Ground Control: "Fine. Everything in order."

Volinov: "Fine, everything fine. We are waiting for contact."

Shatalov: "Approach. Everything normal. Contact! Link-up. Docked!"

Ground Control: "We can see every detail. Everything in order."
Shatalov: "We are turning."
Ground Control: "That is normal. The combination will stabilise itself shortly."
Shatalov: "The spaceships are firmly connected."
Volinov: "Welcome!"
Shatalov: "Welcome Baikal!"
Ground Control: "The spaceships have made a text-book link-up."

At 11.20 the docking probe of "4" slid into the receptor of Soyuz-5. At the same moment the two spacecraft were connected mechanically as well as electrically, and the first experimental space station had become a fact. A "space dwelling", consisting of four compartments had been established, with a total capacity of 18 cubic metres.

During the 35th orbit of Soyuz-4, Yeliseyev and Khrunov began their preparations for a joint excursion into the cosmos. Up to that time they, like their two colleagues, had been dressed in the familiar training suit. They now had to change.

First of all the three Soyuz-5 cosmonauts moved through the connecting hatch into their work compartment. Khrunov and Yeliseyev took two spacesuits from the wardrobe and put them on, helped by commander Volinov. This proved to be none too easy, for the men were weightless and although the two space walkers steadied themselves on the handrails along the walls, Volinov had great difficulty in getting his two comrades into their working gear. But finally the spacesuits and the life-packs which the cosmonauts carried on their backs had been given a thorough check. Khrunov and Yeliseyev shook hands with Volinov and wished him a safe landing. The commander then returned to the command module, the connecting hatch was closed and the work compartment or "dressing room" was depressurised. Then the outer hatch opened slowly. Khrunov was the first to put his helmeted head outside.

"My first impression," so Khrunov said later, "was as if I were standing by the open door of a plane, ready to make a parachute jump, although the height was 200, not 2 kilometres. Then I concentrated on doing everything in the correct order. Once outside I checked the spacecraft, but I found no irregularities. I looked at the earth and then at the sun. Even through the dense lightfilters it is incredibly bright, brighter than if you look at it without visors from the height at which a plane

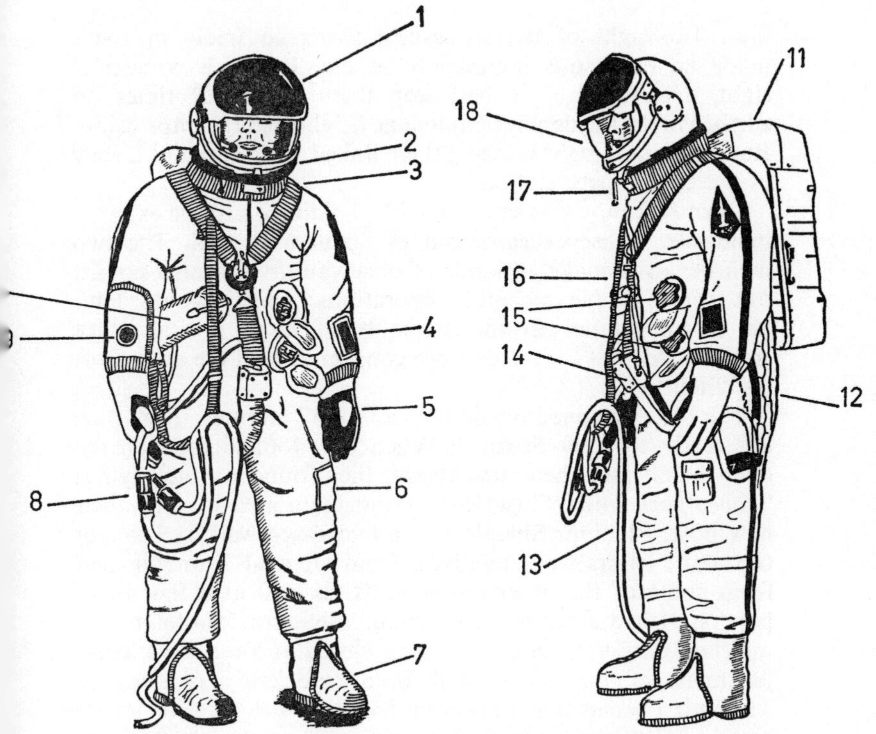

Fig. 17. Spacesuit for extra-vehicular activities (used in the Soyuz project).

1. Hinged sunvisor. Protects eyes and face from intense solar radiation.
2. Helmet attachment ring.
3. Support straps for ASZh (*Avtonomnaya Sistema Zhiznyeobyes-pyecheniya*—"Life Support Pack").
4. Mirror.
5. Gloves, attached to spacesuit by means of bayonet catch.
6. Pocket.
7. Space walking shoes with ridged soles.
8. Safety-line connection.
9. Radiation meter.
10. Connection between "Life Support Pack" and spacesuit.
11. "Life Support Pack", containing sufficient oxygen for two hours; also regulates the temperature of the spacesuit.
12. Reinforced back.
13. Safety-line with built-in telemetry and telephone connections.
14. Clasp of carrier strap.
15. Connection for linking spacesuit to in-board systems (shown connected).
16. Support strap for safety-line.
17. Helmet bolt.
18. Interior visor, hermetically sealed into space helmet.

flies. The sight of the spaceships flying so freely in space made an enormous impression on me. It was a wonderful sight, even though we had seen them dozens of times on earth. Maybe the deployed antennae heightened the impression. But the main thing was seeing them linked. This gave one a view of the entire space station."

Once Khrunov was safely outside, holding on to the external handrails, Yeliseyev came out of Soyuz-5 as well. The two men began to make all kinds of observations and also carried out some simple assembly operations, which would later be of importance in the assembly of large orbital space stations. All this time they were connected with the spacecraft by "lifelines".

The men remained outside for about 37 minutes. They then got ready to enter Soyuz-4. When they looked through the open hatch into their new abode, they found a note saying: "*dobro pozhalovaty!*" (welcome), put there as a joke by their new host, Vladimir Shatalov. The two space walkers brought the mail: letters from relatives, from General Kamanin and from some of the other cosmonauts, as well as a few newspapers. "I had a heavy cosmic mail," Shatalov said later.

When the outer hatch had been closed and the work compartment of the Soyuz had been sufficiently pressurised, Vladimir opened the connecting hatch to welcome his guests and to help them undress. The two men put on new training suits and took their seats. Shatalov treated his cosmic brethren to blackcurrant juice, since vodka is banned in fast traffic!

Shortly after this successful change-over, Soyuz-4 and 5 were once more disconnected. Slowly but surely they moved away from each other. The experimental space station had ceased to exist.

A first-class ship

"It was never intended that the spacecraft would remain linked for a longer peiod," Shatalov said later at the press conference. "The order to separate came after three orbits." He also said that he had acted as "flying instructor". The two scientists Yeliseyev and Khrunov had each in turn piloted the spaceship.

"When it was time to return, the spaceship obediently left its orbit, after the propulsion unit had been switched on," said Shatalov. "At a height of about 10 kilometres the parachutes deployed. The soft landing engines did their job just above the ground and this resulted in an exceptionally soft landing. During our descent I maintained communication with the helicopters and planes of the recovery team." Boris Volinov, the commander of the Soyuz-5, described his spacecraft as "a first-class ship for further exploration of space".

Yevgeni Khrunov, the engineer, said that, as a technician, he had been kept busy. In Soyuz-5 he had tested the docking gear and the mechanical connections of the two spaceships. In addition he had used a sextant to take the altitude of certain stars (an excellent aid to navigation), had studied the way in which radio waves passed through the ionosphere and had carried out many other experiments. Outside the "station", his task had been to test the airlocks, to carry out simple assembly operations, to check the (new) spacesuits and life-packs, to observe celestial bodies and the earth, and to make photographs and cine-films.

"After dozens of flights in a flying laboratory, with short periods of weightlessness, we were well prepared for our tasks and nothing unexpected occurred," said Khrunov. "We could have stayed out much longer," he added.

Why had the Russians entered the second spaceship from the outside and not through the connecting hatch? "There were good reasons for this," said Alexei Yeliseyev. "Such an experiment is of the greatest possible value for the assembly of heavy space stations in the not too distant future, for inspecting the outside surface, and for carrying out repairs and tightening screws in space."*

The spacesuits which the cosmonauts wore were a great improvement on Leonov's. "The suit was comfortable and flexible," said Yeliseyev. "Ventilation and thermostatic control were such that we never felt too hot while we were working outside."

Shatalov said that he had been most impressed by the magnificent view of Soyuz-5 after the two capsules had separated. Initially the crews could see each other's spacecraft only by means of monitors and periscopes, as the view was

*Later, Shatalov told me that it was not possible to move from one spacecraft to another from the inside.

obstructed by the work compartments. After the two ships had been disconnected, Volinov and Shatalov turned their craft 90 degrees. "Through the porthole I could see a brilliantly sparkling ship moving against the background of the far-distant earth," said Shatalov.

Boris Volinov, in answer to a question from a journalist, said that "the time is not far off when scientific space stations will be circling the earth in permanent orbit".

Khrunov made a significant statement: "In the design of our spacesuits certain aspects of Leonov's suit were taken into consideration. Our experiences on this flight may well contribute to the designs of a *moon suit.*" This was by no means the first indication that the Soviets had not lost their long-standing interest in the moon.

Yeliseyev added that assembly work in space in a spacesuit requires a fair amount of extra exertion. The Americans had had the same experience on their Gemini flights. "Work outside the spaceship included installation and dismantling of film cameras and brackets, handrails and television apparatus," said Yeliseyev.

In conclusion Professor Mstislav Keldysh said that a flight during the severe winter season was possible only because of the extreme accuracy with which a guided Soyuz spaceship could be landed and of the efficiency of the recovery team.

"In my opinion," said Keldysh, "it is a question of years, not of decades, before large permanent space stations will be functioning."

Summary

Soyuz-4. *Launched: January 14, 1969, 10.39 Moscow time. Initial weight approx. 6 tons. Angle of orbit with equator: 51·7 degrees. Orbiting time: 88·5 minutes. Perigee: 173 km. Apogee: 224 km. Total duration of flight: 71 hrs. 14 mins. Landed on January 17, 1969 at 09·53 Moscow time.*

Soyuz-5: *Launched: January 15, 1969, 10·14 Moscow time. Initial weight approx. 6 tons. Angle of orbit with equator: 51·7 degrees. Orbiting time: 88·5 mins. Perigee: 200 km, apogee: 230 km. Total duration of flight: 72 hrs. 46 mins. Landed on January 18, 1969 at 11 o'clock Moscow time.*

TRIO IN SPACE

Why no docking? – Every 24 hours lift-off – The trio is complete – Rendezvous – Assembly training – For science and economy

Failure in space link-up?—Doubt concerning construction of Russian space platform—Setback on Soyuz-6 flight?—Russian "space ballet" according to plan—Great uncertainty concerning objective of Soviet space experiment—Soyuz flight fizzles out—Speculations on Soyuz flight.

These are just a few of the newspaper headlines which appeared in mid-October 1969. On the whole the west was certainly none too impressed by the flight of the first cosmic triplets, the Soyuz 6, 7 and 8. Although this was the first time that three manned spacecraft and no fewer than seven cosmonauts were in space simultaneously, such an experiment seemed to be a bit of an anticlimax a few months after the first American moon-landing and only a month before the start of Apollo-12. The flight was certainly not a sensational one, but at the same time there was no question of failure. On the contrary, lengthy conversations in Moscow with General Kamanin and the cosmonaut Shatalov, the commander of the triplet flight, convinced me that the operation had been a significant scientific and technical success. . . .

Every 24 hours a launching

The three Soyuz were launched at a breathtaking tempo: one every 24 hours. On October 11, Soyuz-6 took off with Georgi Shonin and Valery Kubasov on board. Soyuz-7 followed on October 12, carrying the cosmonauts Anatoly Filipsyenko (commander), Vladislav Volkov (flight engineer) and Viktor Gorbatko (engineer). Just under 24 hours later the trio was completed by the start of Soyuz-8. The crew of this spaceship consisted of the veteran cosmonauts Vladimir

Shatalov and Alexei Yeliseyev. Each of the three capsules circled the earth for five days, and consequently they landed at one day intervals—all in the neighbourhood of the town of Karaganda. There was no link-up. At first glance it appeared that the only operation of any importance had been some welding experiments carried out by Valery Kubasov in Soyuz-6.

However, closer investigation shows that the seven men put in a lot of work, mainly connected with preparations for a space station. Moreover, a great many activities took place which proved that manned observation posts in orbit round the earth can be of enormous benefit for the economy of a country as large as the Soviet Union. This may perhaps best be illustrated by a somewhat more detailed report of the week from October 11 to 18, during which the experiment took place.

Soyuz-6

October 11. At ten minutes past two in the afternoon Soyuz-6 lifts off from Baikonur. Tass announces that the flight has several objectives (naturally nothing is said about Soyuz-7 and -8 at this stage):

—to test the systems of an improved version of the Soyuz;
—manual guidance control, orientation and stabilisation of the spacecraft;
—scientific observations and photography of extensive areas of the earth, which may be of geographical or geological value;
—research into the earth's atmosphere for the purpose of applying the knowledge thus gained to agriculture and the national economy;
—bio-medical research into the effect of weightlessness on the human body;
—welding experiments in a vacuum.

In the second orbit, when the large solar panels have been deployed, Shonin orients the spacecraft towards the sun. At eight minutes past eight that evening the course is corrected on instructions from ground control. As soon as the spacecraft has lost direct radio contact with the control centre, the cosmonauts go to sleep in the work compartment.

Soyuz-7

October 12. Soyuz-7 takes off at a quarter past two. At that moment Soyuz-6 is approaching the launching base from the south-west. With the exception of the welding experiments, the crew's tasks are largely the same as those of Soyuz-6's. In addition Tass announces that the two spacecraft will carry out "joint manoeuvres". It will, however, be some time before the crews of the two capsules catch sight of each other. When Soyuz-7 has gone into orbit, the two are dozens of kilometres apart and will remain so for some days. In the fourteenth orbit of Soyuz-6, Shonin carries out some steering manoeuvres, navigating by the stars. Valery Kubasov handles the sextant. When he has completed his measurements and compares the results with the information from ground control, it appears that his calculations are not far out. Later, while Shonin is testing the automatic guidance system, Kubasov is busy recording the landscape passing below. From the portholes of the work compartment he takes numerous photographs of the low-lying coast of the Caspian Sea and the Volga-delta. He also takes pictures of the immense forests of Central Russia and the cloud formations above his socialist fatherland. Together with the photographic material provided by unmanned satellites, these pictures will prove to be of great value to geologists and meteorologists. The results of their studies will benefit the national economy as a whole, as well as agriculture.

The trio is complete

October 13—1.29 p.m. Persistent rumours that a third Soyuz is to be launched prove correct. Soyuz-8, manned by the two veterans Vladimir Shatalov and Alexei Yeliseyev, start chasing their two precursors. Tass announces that Shatalov is in command of the trio.

Only now do the important manoeuvres really begin. Shonin, in command of Soyuz-6, is told which operations he will have to carry out in order to approach the Soyuz-8 which has just been launched. In the 32nd orbit the spaceship is correctly lined up, with the two main engines pointed in the right direction. The engines are then fired and the spacecraft goes into a new orbit. The flight control centre is kept busy: it is no simple matter to control three spacecraft at the same time. Computers instantly calculate the constantly changing trajectories of the trio. Similar actions will be necessary when one day a space station is assembled from various elements, so the operation is of the greatest importance.

October 14. Tass announces that the trio is carrying out joint manoeuvres and observations. The crews are in regular radio contact with each other—for the first time partly via the communication sputnik Molniya-1—but cannot yet see each other, as the distance is still too great. The actual rendezvous (they will remain at a distance of 500 m from each other) is planned for the following day. Shatalov puts his Soyuz through its paces. He carries out numerous manoeuvres, partly by means of the automatic guidance system, partly manually and, like his colleagues, he maintains regular contact with the tracking stations in the Soviet Union. When these are out of reach, for instance when the spaceships are flying over the Atlantic, contact is still possible with the big space communication ship 'Cosmonaut Vladimir Komarov', which is cruising there. A further seven ships of the Academy of Sciences form part of the chain of tracking stations: the 'Morzovyets', the 'Nevyel', the 'Bezhitsa', the 'Dolinsk', the 'Ristna', the 'Kegosyov' and the 'Borovitsyi'. Compared with the Americans, the Russians were for a long time at a disadvantage, because they had tracking stations only on their own soil. The fact that they are now using special ships, equipped with the most advanced electronic apparatus, proves that they are determined to be well prepared for future large-scale space operations, in which good communications will constitute the difference between success and failure.

Rendezvous

October 15, the day of the rendezvous. Extensive manoeuvres are carried out, using manual as well as automatic systems. The computers in the ground control centre are running to capacity. Presently, when all the necessary manoeuvres have been completed, the orbits of the three Soyuz craft should be identical. But today's programme comprises more than this.

Between course corrections, the crew of Soyuz-6 is carrying out all sorts of medical tests, including some in connection with the way the vestibular mechanism of the inner ear functions in conditions of weightlessness. Kubasov, moreover, continues to take numerous photographs of various interesting areas of the world and of whirlwinds. Filiptsyenko, Volkov and Gorbatko concentrate their efforts on the stars; they also take photographs. Shatalov and Yeliseyev in Soyuz-8 are chiefly occupied with course corrections. Shatalov, the commander of the "cosmic squadron", as the Russian press later calls the trio, is moreover busy with the task of controlling the three spaceships. It is he, after all, who is responsible for the rendezvous operations.

Finally the three spaceships are correctly aligned in an orbit with a perigee of 200 km and an apogee of 225 km, at an angle with the equator of 51·7 degrees. At this moment the distance between Soyuz-7 and Soyuz-8 is 500 m, while Soyuz-6 is a little further away. The cosmonauts of "7" and "8" can see each other's spacecraft. There will be no docking; this had been done before, in January.

October 16. The distances between the spaceships have once more increased. In Soyuz-6 Kubasov is experimenting with welding, as announced. These experiments are carried out with automatic apparatus in the work compartment, which has previously been depressurised. Kubasov conducts the process from the command module. He uses three different welding techniques: plasma arc machining, arc-welding (with electrode) and electron beam welding.

The experiments are completely successful. This kind of work will be of importance when large orbital space stations are being assembled. When the experiments are concluded, the work compartment is once more pressurised and Kubasov enters to gather up the results.

This concludes the task of Soyuz-6 and its crew. That afternoon at eight minutes to one, the capsule lands 180 km north-west of Karaganda.

October 17. At 12.26 Soyuz-7 lands not far from the point where the "6" came down the day before: at 155 km north-west of Karaganda. Shatalov and Yeliseyev in Soyuz-8 carry on for a time with medical experiments. They also take some more cine-films of areas of geological interest on earth. Their library of photographs and films is by now of impressive size.

October 18. At 12.10 Soyuz-8 also makes a perfect and very soft landing in the snow, 145 km north of Karaganda.

Why no docking?

Immediately after the flight the West persistently asks: "Why was there no docking?" Had something gone wrong? Why all those course corrections if it was never intended to link up at least two of the three spacecraft?

Shortly before the flight of Soyuz-9 I met Shatalov in Moscow. The first question I asked him was:

"*What do you think were the most important results of the flight?*"

He answered rather evasively: "There were several important results."

"*But which was the most important?*"

Shatalov: "That is difficult to say. It is difficult to distinguish between what was very important and what less so. By the same token it was difficult to say of the flight of Soyuz-4 and -5 which was the more important: the transfer from one spaceship to the other, or the link-up. Without link-up there would have been no transfer."

"*But on the flight of the '4' and '5' there were at least two important items on the programme: docking and transfer. Which then were the two main points of the Soyuz-6, -7 and -8 programme?*"

Shatalov: "To begin with there was the mere fact that there were three capsules in space at one and the same time. One of the American astronauts once said that the US would not be capable of such an operation. Then there were the experimental manoeuvres. There were thirty approach manoeuvres, that is quite something: we have learnt to find each other, to link up* and to move apart again. Moreover the welding experiments were of immense significance for the construction of future space stations."

"*But what is the essential difference between a double and a triple flight?*"

Shatalov: "Guidance control is much more complicated. We

*Here Shatalov referred to Soyuz-4 and -5.

had to steer and control three spaceships simultaneously. I recently read an article—I think it was in the magazine *America*—by either Lovell or Armstrong. They said that it might be difficult for ground control to keep two spacecraft in hand at the same time—the lunar module and the command module. In the final analysis it is a question of bringing several elements together in space, and for that reason we have studied in how far it was possible to bring three capsules together and keep them there."

"*But they did not dock?*"

Shatalov: "No, we did not. It was not the most important problem. Only bringing them together was a difficult operation."

I wanted to draw him out further on this question of docking.

"*I understand that Soyuz-6 had no docking mechanism. Is that true?*"

Shatalov: "Yes. It was officially announced before the flight."

"*Does that mean that Soyuz-7 and -8 did have such a mechanism?*"

"No, Soyuz-7 and -8 were not equipped for docking either. It was not necessary; we had already done this in January. And before that it had been done, entirely by automatic control, with Kosmos satellites."

So it was clear: as the Soyuz craft *could* not link up, such an operation was never intended. The rumours spread in the West, that perhaps a central section should have been launched into which the Soyuz craft could have docked, were therefore entirely unfounded. In any case it would have been illogical to launch the manned spaceships first and then such an unmanned coupling unit, a kind of rudimentary space station, with a possibility of failure.

Shatalov was right: on this occasion there was no sense in linking the three Soyuz. Subsequently it became obvious that the Soyuz was not intended to form a "construction unit" in the building of a space station. It was primarily a transporter of men and materials. The docking of Soyuz-4 and -5 was of significance as an experiment only. What was important was the manoeuvring with different spaceships. At a future date a large space station would be sent up in sections. These sections would have to orbit close together for hours, if not days, before cosmonaut technicians could assemble them.

For science and economy

In other respects, too, the work of the three crews—although not spectacular—was of immense value, namely for the national economy and for science.

There were in the first place the geological observations and the photographing of objects which might be of value to geologists. The question was studied of how far it was possible to determine from an orbit round the earth where certain minerals could be found. The reflective capacity of various types of landscape (forests, deserts, mountains) were also studied. These observations were of all the greater value since the same objects could now be seen simultaneously from three different positions. With ordinary aerial photography it is not possible to map out large areas of the earth's surface at once. This could now be done.

The meteorological observations and the photographs of cloud formations were also of immense importance. Shonin, for instance, tracked a moving cloud formation over Kazakhstan and reported a typhoon off the coast of Mexico. He also discovered a particularly strong whirlwind over the steppes of Kazakhstan, not far from the spot where he would subsequently have to land his Soyuz-6. His observations indicated that the wind in the landing area would abate, so that it was not necessary to find another landing place. In this way the crew of the "6" contributed to their own safe landing.

The crews of the three Soyuz also traced a hurricane west of Mexico and two tropical storms in the Atlantic and in the Indian Ocean.

All seven made numerous astronomical observations, such as measuring the brilliance of stars and of the sun.

Shortly after the flight the chief designer of the Soyuz commented on the experiments in welding which Kubasov had carried out.

"Some people," he said, "believe that space stations will be assembled entirely automatically, but this is not the case. In my opinion space stations will, as far as possible, be assembled automatically, but manual operations will always be involved as well. One of these manual operations will be welding."

SOYUZ-6, -7 and -8 FLIGHTS IN FIGURES

	Launched	*Immediately after launching*				*At time of Soyuz-7 launching*				*After manoeuvre October 15*				*After manoeuvre October 17*				Landed
		Apogee	Perigee	Angle with equator	Orbiting time	Apogee	Perigee	Angle with equator	Orbiting time	Apogee	Perigee	Angle with equator	Orbiting time	Apogee	Perigee	Angle with equator	Orbiting time	
Soyuz-6	Oct. 11 14 hrs. 10 mins.	223	186	51·7	88·36	230	194	51·7	88·6	225	200	51·7	88·6	—	—	—	—	Oct. 16 12 hrs. 52 mins.
Soyuz-7	Oct. 12 13 hrs. 45 mins.	226	207	51·7	88·6	226	207	51·7	88·6	225	200	51·7	88·6	—	—	—	—	Oct. 17 12 hrs. 26 mins.
Soyuz-8	Oct. 13 13 hrs. 29 mins.	223	205	51·7	88·6	—	—	—	—	225	200	51·7	88·6	256	190	51·7	88·95	Oct. 18 12 hrs. 10 mins.

MARATHON IN SPACE

Surprise for Neil Armstrong – General Kamanin: flights of more than fifteen days are difficult – Physical jerks in Soyuz-9 – Seventeen days and seven hours in space – Manoeuvring exercises – Problems of adaptation

June 1, 1970. In Moscow it is a little past ten in the evening. Somewhat further north, in Zvyezdnii Gorodok (Star Village), Neil Armstrong, the first man on the moon, is the guest of cosmonaut Georgi Beregovoi. Neil has come to the Soviet Union to attend a congress in Leningrad. At the same time he is taking the opportunity to present Prime Minister Kosygin with a piece of moonrock and to give commemorative medals of Gagarin and Komarov to their widows. Similar medals were taken to the moon a year before and left at "Tranquillity Base" by the lunar module Eagle.

The hands of the clock are moving towards 10.30. At that moment Beregovoi switches on the television. Almost immediately a slender white rocket appears on the screen, sharply contrasted against the black night sky. "This is in your honour," Beregovoi tells the surprised Armstrong with a straight face when a few seconds later the rocket leaves the launching pad to disappear into the dark sky. The cosmonauts Andrian Nikolayev (Valentina Tereshkova's husband) and Valery Sevastyanov (a newcomer to space travel) are on their way in Soyuz-9.

As usual speculations are rife. Is a start at last being made with the long-awaited space station? No, Tass squashes all rumours by announcing in the very first communiqué that it is a "solo-flight" in an orbit close to earth. Surely it will be a very special flight? And indeed: Soyuz-9 does not land until June 19, exactly seventeen days and seven hours after it was launched. Nikolayev and Sevastyanov have thoroughly beaten the duration record of fourteen days established by the Americans Frank Borman and James Lovell in Gemini-7. Even before landing they receive a telegram from their American colleagues congratulating them on their achievement.

Kamanin: the problem of weightlessness

The main objective of the Soyuz-9 flight was to find out how long man could remain weightless without ill effects. This, of course, was of immense importance, for in future cosmonauts would have to live in a space station for weeks, if not months, at a time.

Less than a month before the start of Soyuz-9 I was in Moscow, where I had a conversation with General Kamanin. On that occasion he indicated clearly that the Russians regard weightlessness as a great problem. At that time I was struck by the fact that the general showed an almost touching degree of good sportsmanship when he remarked that the Americans had established a record with their fourteen-day flight.

"And this has never yet been beaten," he said, "although it was established nearly five years ago. We believe that man will have difficulty in standing up to very much more prolonged spaceflights without special aids. In other words, on such flights—for instance in space stations—we shall have to provide artificial gravity."

I asked the general: "Do you think anything could be achieved with further training?"

Kamanin: "No, I don't think so. It is possible that man would eventually adapt after living in space for a million years. If he had been born in weightlessness and lived and worked in it. Perhaps. But for earth-dwellers it remains a problem."

General Kamanin's remarks were probably partly inspired by the experiences obtained with the Kosmos-110 dogs, which orbited the earth in February and March of 1966 for three weeks at a stretch. Nevertheless the conclusions drawn from this operation could not be unconditionally applied to cosmonauts, since the dogs had little opportunity to move in Kosmos-110, whereas astronauts on board a spaceship can do a great deal to keep in condition with special physical exercises. Exercising is therefore thought to have been one of the most important activities on the flight of the "9".

The most worrying symptom which had presented itself during the Kosmos-110 flight was the absorption of bone calcium by blood and urine. After the flight the dogs' bones were considerably weakened for in conditions of weightlessness the frame loses its supporting function. A prolonged stay in bed may also result in loss of calcium, but the Kosmos-110 flight was a revelation in this respect.

"What," Russian space doctors asked themselves, "will become of future space travellers who stay for weeks, if not months, in a space station—at any rate if no artificial gravity is provided? Will they be able to withstand the deceleration process on their return into the atmosphere? Will there be any permanent damage to their system?"

A lot of exercise

Nikolayev's and Sevastyanov's experiences of weightlessness during their marathon flight were not unfavourable. Day after day Tass reported that the cosmonauts were still feeling fine, that they continued their flight with enthusiasm, and above all, that they remained capable of a great deal of work. Space-flights are no longer purely experimental operations. As the flights become more and more routine exercises, the crew is expected to "earn its living". Hence the accent, during the long Soyuz-9 flight, lay chiefly on useful scientific and technical work, which could be of great value to the economic progress of the socialist fatherland. For the first time the emphasis lay clearly on the economic significance of the flight—more about which anon.

It soon became obvious that, as far as the crew's health was concerned, regular exercises were considered of the greatest importance. Nikolayev and Sevastyanov were instructed on no account ever to neglect them. This was the only way in which they could prevent, as far as possible, the weakening of their muscles as a result of the continued absence of gravity.

But naturally the cosmonauts' good health during the flight was not enough in itself. The space doctors were particularly worried about their condition during and after their return to earth. On their re-entry through the atmosphere the cosmonauts would be subjected on average to three times their own weight, while once back on earth they would have to adapt again to normal gravity.

At 14.59 Moscow time on June 19, the command module of the Soyuz-9 landed in the fields of the Collective "Lenin" near the village of Intumak, 75 km west of Karaganda. The two cosmonauts were so weak that they were unable to climb out of the capsule. Nevertheless they ate their first meal on earth with a good appetite.

A detailed medical examination, taking several days, was to show that they were not yet themselves. A space doctor declared frankly that it was "no simple matter" for a human body to re-adapt itself to gravity after sustained weightlessness.

What had happened? A number of provisional conclusions could be drawn from a report issued by Tass less than a week after the landing.

By June 23 the cosmonauts' blood pressure was still not normal, which indicated that their bodies had difficulty in adapting to conditions on earth. The cosmonauts themselves confirmed this: they were not yet feeling in top condition. According to Tass this led the medical men to conclude that "it was quite probable that it would be a good thing to produce artificial gravity in a space station, for it appears that it is no easy matter for man to re-adapt to terrestrial conditions after a prolonged stay in space".

Tass did not mention the report which had previously appeared in the army paper *Red Star*, that the cosmonauts' eyesight had deteriorated somewhat during the flight. Their ability to distinguish between colours (Nikolayev and Sevastyanov underwent a number of tests in this respect) also appeared to have been seriously affected by weightlessness. "It was noted that the reliability of visual observation had already considerably worsened 24 hours after lift-off," *Red Star* wrote on June 16. "Scientists argue that the co-ordination of the muscles which control eye movements is affected by weightlessness." In particular, it was the bright colours which Nikolayev and Sevastyanov could no longer see so well. "The greatest deviations occur when looking at purple, pale blue and green," wrote *Krasnaya Zvyesda*. "Lesser deviations occur when looking at red. This proves that certain deviations occur in the mechanism which distinguishes between colours. Medical science has yet to discover the cause."

On June 27—over a week after the two Soyuz-9 travellers had returned—the cosmonaut doctor Boris Yegorov wrote a detailed article in *Izvyestia*, in which he asserted that, with good training (particularly during the flight!) a stay of a month in weightless conditions was not impossible.

Towards the time when cosmonauts will remain in space for

longer than a month it would be necessary to launch space stations which revolved on their centre of gravity in such a way as to create artificial gravity.

Manoeuvres

Naturally the biological aspects of the Soyuz-9 flight were the most important. Nevertheless Nikolayev and Sevastyanov made good use of their nearly-18-day voyage by carrying out all sorts of scientific operations.

"Not only the men, but also the technical systems on board were tested as to how they would last the course," Nikolayev wrote in a report of his experiences. "We tried out new systems for navigation and guidance control. Manoeuvring in the cosmos is of immense importance for the construction of space stations. Consequently we devoted a great deal of time to course corrections."

In these course corrections the steering rockets and the main rockets of the Soyuz naturally played a large part. For accurate correction it is essential not only that the main engines develop the correct amount of thrust for a precisely calculated number of seconds, but also that this thrust is developed in the right direction. To begin with, therefore, the spaceship must be placed in the correct position. This is achieved by means of small steering rockets attached to the three sections of the spaceship. Nikolayev gave an example of such a manoeuvre:

"While the spaceship was flying in the earth's shadow, we orientated ourselves on the star Vega.* By manual control and with the aid of the optical visor in front of us, we brought the star within the field of vision of our stellar homing device.† The star-finder caught the star, after which the spaceship was stabilised in the relevant position with the aid of gyroscopes. We then checked how accurately the gyroscopes had functioned."

Nikolayev added that the new stellar homing device (used for the first time on this occasion) had proved to be very accurate, even when the spaceship was not flying on the dark

*In the constellation of Lyra.

†An electronic homing device, such as is also used for the orientation of unmanned probes.

side of the earth, when other strong sources of light (the sun and the earth) interfered. Nikolayev and his colleague carried out several tests to determine whether it would be possible to navigate their Soyuz entirely independently of ground control. In future, when spaceflights become a matter of everyday routine, this will be increasingly necessary. In addition it is, of course, essential when for some reason or other the much-appreciated radio link with earth is lost.

Nikolayev said: "We measured the orbiting time of our spaceship, the height of its trajectory at various points, the angle with regard to various beacons on earth. With the aid of a new type of sextant we took the altitude of stars, as against the earth's horizon, by day and by night. We could thus determine the spaceship's trajectory and calculate which corrections had to be made. In doing so we of course made extensive use of the in-board computer."

Nikolayev related that the course-corrections were carried out manually as well as automatically. "The spaceship could be oriented with great accuracy, even when manually controlled," said the space veteran. "And what is more," he added, "with minimum fuel consumption!" Since every pound of initial weight counts, this, of course, was of great importance.

Sevastyanov, the engineer, devoted a great deal of his time to checking the construction of the Soyuz: deformation, temperatures resulting from one-sided solar heating, hermetic sealing, etc.

"Masses of photographs were taken," said Nikolayev. "These will be of value to geologists, oceanologists and meteorologists. Not only black-and-white photographs, but also colour slides and photographs within a specific band of the spectrum.* This makes it possible to produce geological maps to be used in the search for minerals and other soil content."

Two sets of underwear

The crew of Soyuz-9 had no cause for complaint about the meals, said Nikolayev. Everything possible had been done to

*Photographs taken in one single colour (e.g. infrared) sometimes show more details than normal plates.

make life as pleasant as possible for the marathon flyers. The cosmonauts not only had hot meals—heated in a small electric oven in the work compartment—but even hot coffee for breakfast. Hygiene on the flight naturally presented problems. The crew adhered to the somewhat old-fashioned principle—which, however, was quite revolutionary as far as space travel was concerned—that one should change one's underwear once a week. Consequently Nikolayev and Sevastyanov had each brought two sets of underwear and two pairs of socks. They could (as yet) not take a bath, but the capsule carried carefully packed wet and dry towels, which had to suffice.

In his report on the flight Nikolayev explained that the Soyuz-9 was the first spaceship to be launched at night. As usual, so-called "experts" in the West had suspected all sorts of reasons for this, but the actual explanation was much more obvious. In the first place both the descent to earth and the actual landing had to take place by day and this meant that, in view of the number of orbits which had been planned for this record flight, the lift-off would have to take place at night. Moreover, a number of experiments over Soviet territory had been planned for specific times, experiments in which ground installations were to play a part. Naturally this meant that the space travellers would sleep and wake at unaccustomed times, but this in itself provided interesting new information from a biological point of view.

The flight even allowed for relaxation and the cosmonauts had in fact one whole day off duty. They used this mainly for playing a long-distance game of chess with General Kamanin and the cosmonaut Gorbatko, which ended in a draw.

Artificial gravity on prolonged flights

And what did Nikolayev have to say about the most important question: the effect of prolonged weightlessness on the human organism? Briefly his opinion was as follows: "Muscles, heart and circulation in particular are prone to deterioration during prolonged weightlessness. Of course we did not wait until this occurred: we did something to counteract it. Twice a day we exercised extensively, using special 'pressure suits'. These could be attached to the floor of the Soyuz-9 by means of rubber bands, so that we could do press-ups and other exercises, in which we had to overcome forces of several dozen kilograms. We also used a ten-kilogram expander which we had to stretch thirty times a minute."

All this contributed greatly to keeping the cosmonauts in reasonably good condition, and to their working and sleeping well. At one time one of them slept so deeply that he hit his head on the ceiling before realising that, under pressure of his breathing, he had floated out of his sleeping-bag!

Nikolayev admitted that for the first three or four days he and Sevastyanov had had difficulty in adapting to their new conditions. After that, and right up to the landing, the men had felt fine. Adaptation to earth conditions after their return was disappointing. After 18 days of weightlessness it was as if their limbs weighed twice as much as before the flight. They found it hard to get up from a chair and thought their beds unusually hard. But all these unpleasant effects disappeared after five or six days, and they then felt as well as before.

Nikolayev's conclusion was as follows: "In principle, prolonged spaceflights present no problems. We now know that the barrier of weightlessness can be overcome. No radical measures are necessary for short flights of one or two weeks. On flights lasting some months in zero-gravity we shall have to introduce strict programmes for providing artificial stress for muscles, heart and circulation. Certain drugs will help to maintain the organism. On very long flights—for instance interplanetary flights—we shall have to resort to artificial gravity, which will also be necessary to prevent serious problems of adaptation when cosmonauts land on other planets."

Nikolayev's conclusion was marked by one striking omission: he never mentioned artificial gravity in orbiting space stations. This led one to suspect that the Russians were not planning to create artificial gravity in their first space station, but this would mean that the crew of such a station would have to be replaced after relatively short shifts. . . .

Soyuz-9 in a nutshell

Launched on June 1, 1970, at 22.00 Moscow time. Initial weight approx. 6·5 tons. Angle of orbit with equator: 51·7 degrees. Orbiting time: 88·59 minutes. Perigee: 207 km, apogee: 220 km. Total duration of flight: 424 hrs. 59 mins. Landed on June 19, 1970, at 14 hrs. 59 mins. Moscow time.

ROBOTS SURVEY THE MOON

Luna-16 returns with moon rocks – The outward voyage – On the moon – Back to earth – Moon soil on earth – Lunokhod-1: the first car on the moon – Luna-18, 19, 20, 21 – One "first" after another

July 20, 1969, 16 hrs. 18 mins. Eastern Daylight Time. It is an important moment in the history of mankind: two Americans land on the moon. One of their tasks is to pick up samples of moon soil and to bring them back to earth.

September 20, 1970, 8 hrs. 18 mins. Moscow time. Another historic moment: a Russian robot lands on the moon. One of its tasks is also to pick up samples of lunar soil and to bring them back to earth.

Exactly fourteen months after the Americans, the Russians, too, had got hold of some moon material. Once more the old question was raised as to whether it was better to use men or robots for the exploration of the moon. A question which became even more topical when the Russian moon-car Lunokhod-1, delivered by Luna-17, started moving on the moon. It is not possible to settle this question by the assertion that unmanned moon probes are less risky.

It was in any case certain that the incredible achievement of the unmanned Luna-16 was technically comparable with the grandiose flight of Apollo-11 and its crew, even though the Russian success did not make such a great impact as Armstrong's and Aldrin's first steps on the moon. Twentieth-century man rapidly becomes accustomed to immense technical achievements.

The outward voyage

Baikonur, September 12, 1970, 16 hrs. 26 mins. Safely concealed under the nose-cone of a huge rocket, Luna-16 leaves the earth. The spaceship consists of two sections: a descent stage and an ascent stage. The descent stage is to be used for making the necessary course-corrections and for the moon-landing. The ascent stage will have to bring the return capsule attached to it safely back to earth together with samples of lunar soil.

Tass announces that the spacecraft has been launched by a more powerful rocket than that used for lifting off Luna-9 and 13, two unmanned space capsules which have previously made soft landings on the moon.

Within a few minutes after the start, the final stage section and the moon scout arrive in a parking orbit round the earth at a height of 212·2 km. The angle of the orbit with the equator is 51 degrees 36 minutes. The Co-ordination and Computing Centre gauges the trajectory and calculates to a split second the moment when the third stage engine will have to be fired once more.

Seventy minutes after blast-off. The moment has come. The third stage functions again and increases the velocity from 8 to 11 km a second. The nose-cone has meanwhile been jettisoned. The Luna is detached from its carrier rocket and now leads an independent existence. It revolves on its axes; it orients itself by the sun and the earth; it stabilises itself.

September 13. Two course corrections have been planned. One of these is now carried out. The Co-ordination and Computing Centre works out extent and direction of the correction and feeds the necessary information into the electronic brain of the moon scout. Again the Luna revolves on its axes, causing the descent stage engine to point in the right direction. Then comes the signal for the engine to fire. It functions for exactly 6·4 seconds and proves to have done the trick. No second correction is necessary. A deviation of 1 metre per second in the new speed would have resulted in a 300 km deviation near the moon! Luna speeds towards the point where four days later the moon will be.

September 17. The crucial moment approaches. At 2.38 Moscow time the descent engine is fired. The moon spy's velocity diminishes, and it comes within the moon's sphere of influence, arriving in a satellite orbit at a height of 110 km. In the course of the next three days there are two further course corrections.

The first correction decreases the periselenium* to 15 km.

*Point of trajectory nearest to the moon's surface. The highest point of the orbit is called aposelenium.

The result is an elliptical orbit at between 110 and 15 km. The lowest point lies near the spot in the Sea of Fertility, on the "left cheek" of the moon's face, where the robot is to carry out its digging operations. The second course correction results in a small change in the plane of the orbit. The periselenium remains at 15 km, but the aposelenium is now 106 km.

September 20, 6.06 Moscow time. The signals for the descent have been fed into Luna's computer. The operations begin. Between 6.41 and 7.31 there will be no radio contact with the Luna, as it will then be behind the moon. The probe will now have to control its own descent procedure.

8.12. The descent stage, now pointing away from the moon, functions once more. Near the periselenium the Luna leaves its elliptical trajectory. Since a radio signal takes just over a second to cover the distance between earth and moon, the Co-ordination and Computing Centre can no longer assist, for in one second Luna covers 2 km. The robot is now on its own. The radio-altimeter goes into action; it sends out signals which rebound off the moon's surface back to the capsule's receiver. From the time taken by this, the on-board computer calculates in a flash the capsule's height above the surface. The engine has stopped. Luna now falls in a wide curve towards the moon. At a height of 16 km the main retro-rocket is fired once again. Height, horizontal velocity and vertical velocity are measured continuously and the engine's thrust is adjusted accordingly. At a height of 20 metres above the surface the speed has been reduced to only 9 km an hour.

The probe switches off its main retro-rocket. Simultaneously two smaller engines are fired, which in their turn are switched off by the altimeter just before Luna reaches the moon's surface.

Eight Hours 18 minutes 20 seconds. The four short legs of the Luna, ending in discs, touch the moon's soil. Luna-16 has landed practically at the exact calculated point, at 0 degrees 41 minutes South latitude, 50 degrees 18 minutes East longitude.

On the moon

The Centre of Remote Space Communications (*Tsenter Dalyneisii Kosmitsheski Svyazi*) in the Crimea switches on the

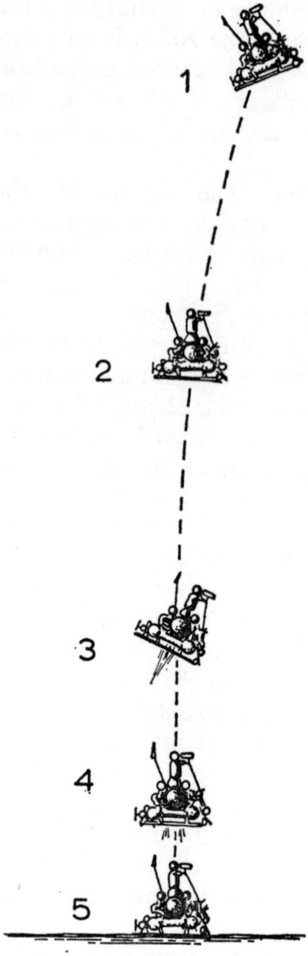

Fig. 18. Landing of Luna-16.

1. Braking operation finished. Luna in descent.
2. Radio-altimeter and speedometer switched on.
3. Start of deceleration stage. Main engine fired. Descent becomes vertical.
4. Main engine cut. Smaller engines continue deceleration.
5. Luna-16 has landed.

Fig. 19. Luna-16. (*right*).

1. Return capsule.
2. Steel bands connecting capsule to ascent stage.
3. Socket of core-drill arm.
4. Return capsule parachute.
5. Lunar soil container with sealing cover.
6. Heat shield.
7. Transmitter.
8. Batteries.
9. Fuel tank for ascent stage.
10. Steering engine.
11. Steering engine.
12. Connection between ascent stage and descent stage.
13. Pressure chamber for steering engines.
14. Attitude control jets, descent stage.
15. Telephotocameras.
16. Bracket for core drilling mechanism.
17. Core drilling mechanism.
18. Swivel mechanism for bracket.
19. Steering engines pressure chambers.
20. Descent stage apparatus.
21. Main descent stage engine.
22. Landing pad.
23. Fuel tank.
24. Steering engine.
25. Detachable fuel tanks, jettisoned before landing.
26. Auxiliary landing engines.
27. Steering engine.
28. Landing structure.
29. Steering engines.
30. Descent stage apparatus.
31. Detachable fuel tank, see 25.
32. Cable sheath.
33. Main ascent stage engine.
34. Steering engine.
35. Antenna.
36. Fuel tank ascent stage.
37. Fuel tank ascent stage.
38. Ascent stage steering engine.
39. Ascent stage.
40. Antennae.

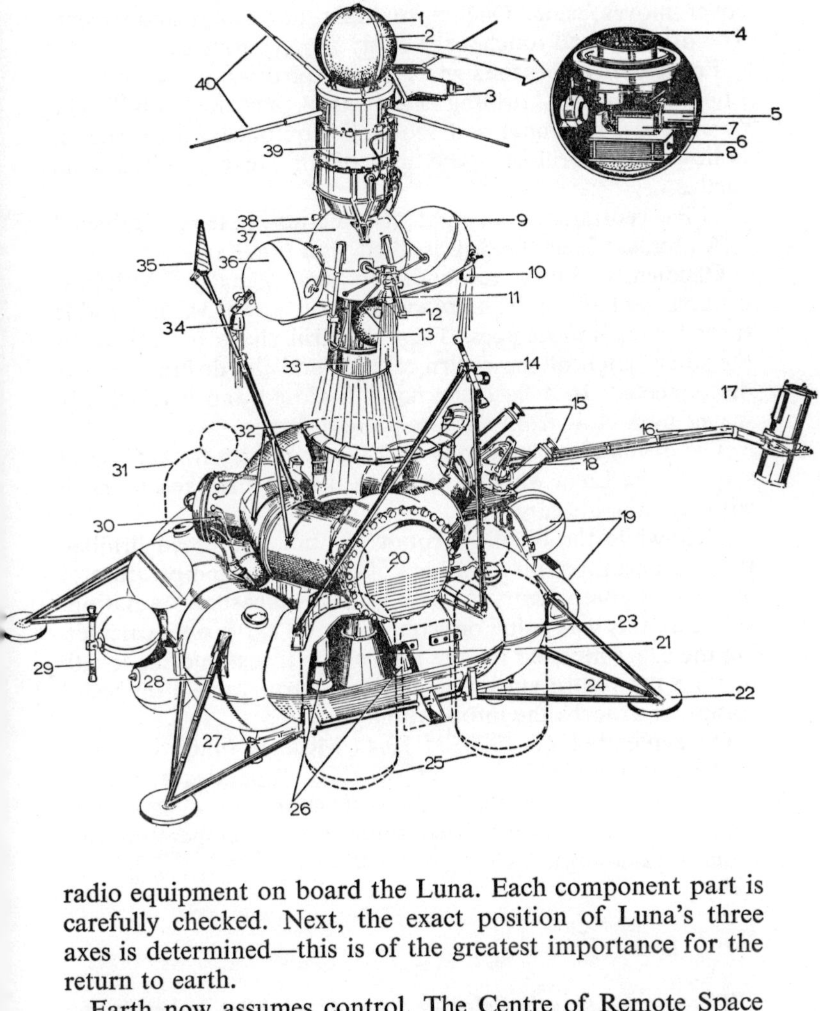

radio equipment on board the Luna. Each component part is carefully checked. Next, the exact position of Luna's three axes is determined—this is of the greatest importance for the return to earth.

Earth now assumes control. The Centre of Remote Space Communications gives the instruction to start drilling. During the flight the drilling apparatus has been tightly bolted down. It is now unlocked. Slowly the bracket and the drilling head swing out. A signal from earth operates the telephoto cameras which are to assist the scientists in finding the best place to drill. The drilling head is turned 180 degrees, so that the drill now points towards the moon's surface. A small protective cover moves aside. One more slight horizontal adjustment. The drilling head touches the dusty moon surface.

From earth comes the signal: "switch on drilling mechanism." Luna confirms: "drilling mechanism switched on." The *Gruntozabornik* (lunar soil scoop) is equipped with a special hollow drill. Drilling speed and depth are controlled from earth.

"Five centimetres . . . eight centimetres . . . ten . . . fifteen." A moment later the drill is filled with lunar soil.

Obediently Luna responds to the signal: "Withdraw drilling mechanism." The bracket moves upwards, the head is turned back 180 degrees. Then the drill slides back through the open hatch of the return capsule and the drilling head is disconnected. In a flash the hatch is closed and hermetically sealed by a blow from an automatic hammer.

The drilling bracket drops back to make way for the ascent stage of the Luna which is to bring the capsule back to earth with its valuable cargo.

Meanwhile the obedient robot has not only been drilling. It has transmitted information concerning the temperature of its various components to earth and has measured radiation. While all this was going on, computers on earth were calculating the exact moment for the return, the thrust and the length of time the return stage engine must burn. Luna's on-board computer absorbs the information.

On September 21, 1970, at 10.43 Moscow time, the signal for the start is given. The return stage separates from the descent stage, using the landing structure as a launching platform. This section will continue to take temperature and radiation readings.

Back to earth

The ascent stage's velocity mounts rapidly. When it has reached 2708 metres per second (a few hundred metres more than escape velocity), the engine is stopped. Luna-16 is now in a ballistic trajectory towards the earth. Its course is so perfect that further correction is unnecessary. On the basis of measurements obtained with huge radio-telescopes, the Co-ordination and Computing Centre can accurately forecast where in Kazakhstan the capsule will land.

September 24, 4.50 Moscow time. At a signal from earth the steel bands connecting the capsule to the engine section recoil. The capsule's own batteries take over.

8.10 Moscow time. The spherical capsule reaches the upper layers of the atmosphere with a speed of just over 11 km per second. The g-forces reach no less than 350, for unlike the Zond and the Soyuz capsules, the spherical vehicle provides no lift. A special damper in the bottom of the sphere cushions the instruments against high g-forces. The temperature of the heat-shield meanwhile has increased to 10 000 degrees Celsius.

The rest of the flight is controlled by the capsule's on-board computer. When maximum temperature and pressure is reached, the gravity-meter gives the signal for the parachute hatch to open. At a height of 14·5 km and a speed of 300 m per second, the drogue parachute is ejected. At a signal from the barometer, the drogue chute is disconnected at a height of 11 km, and the main parachute is deployed. The barometer also gives the signal for switching on the radar equipment which operates on the compact on-board batteries.

8.14 Moscow time. Planes and helicopters patrolling the recovery area receive the capsule's radio signals. A moment later one of the helicopter pilots catches sight of the capsule, swinging down under its parachute.

8.26. The blackened capsule reaches Soviet soil with a gentle impact, landing in farmland 80 km south-east of the town of Dzhezkagan. People living nearby rush to the spot. An observer reports that the capsule is in good condition. It is immediately put on board a helicopter and flown to Moscow, where the container holding the lunar material is handed over to the Academy of Sciences.

Moon soil on earth

The hermetically sealed container removed from the capsule is taken to a specially equipped laboratory of the Academy of Sciences. Before it is deposited in the "reception chamber", it is measured for radiation and its outer surface is thoroughly sterilised. (This had also been done before the start.) The reception chamber is equipped with mechanical arms and with apparatus for removing the drill containing the lunar soil. The sample is weighed and studied in this chamber. Subsequently samples will be transferred to various specialised laboratories.

The chamber is depressurised as soon as the container has been deposited, to prevent the lunar material being contaminated with elements of the earth's atmosphere. As an added precaution, the chamber is filled with helium, which does not react with other elements. To prevent air from seeping in from outside, the chamber is overpressurised. These precautionary measures have a double purpose: they are intended to protect terrestrial organisms against contamination by lunar material, and vice versa. Similar measures were taken in America with the samples brought back by the Apollo crew.

Sterile mechanical arms, operated by a specialist, open the container and take out the drill with its lunar content. The drill is covered by a thin layer of grey dust. The steel arms are operated with the utmost care. Gently they transfer the drill's content to a transparent bowl, taking care to maintain its shape. In this way an idea may be obtained of a cross-section of the moon's surface at the place where Luna-16 landed. Now the sample can be extensively studied and photographed through optical glass portholes. Colour photographs are made with various forms of lighting and from all angles.

The major part of the core consists of fine-grained particles, grey in colour. Its appearance shows that it is porous in character and has great powers of adhesion. In this respect the "Russian" lunar soil does not differ from the "American".

And what about gamma radiation? Scientists conclude that this is no more intensive than the natural radioactive radiation of certain terrestrial kinds of soil.

Next, the samples are subjected to toxicological and biological examination. The Russian scientists, like the Americans, conclude that there is no question of "lunar bacteria". Not until the biological tests have been concluded are the samples released for examination in specialised laboratories, where they undergo radiation tests, chemical, physico-mechanical, thermophysical and other tests. This research, according to Tass, will teach us a great deal, not only about the structure and origin of the moon, but also of the earth. The day will come, says Academician Petrov, when we shall be able to obtain similar samples from the planets, which will contribute further towards completing our knowledge of the creation of the solar system. . . .

It was as early as October that the Soviets fulfilled their promise to publicise the results of their research. At a press conference the Soviet scientist A. P. Vinogradov gave a survey of the results obtained so far. Briefly these were as follows: Luna-16 landed 100 km west of the crater Webb in the Sea of Fertility, one of the relatively level areas of the moon. This is 900 km from the place where the Apollo-11 samples were taken. After having cut easily through the powdery surface material, the drill penetrated only 5 mm into the basalt-like layer underneath. Consequently only the end section of the column (35 cm long) was coarse-grained. In the main, the colour of the Luna-16 sample, which weighed approx. 100 g., is dark grey. The material looks sticky; it adheres to the sides of the bowl. In spite of the total absence of water, the lunar soil resembles moist sand. Its average specific gravity is 1·2 g per cubic centimetre.

In principle two categories of particles can be distinguished:

—particles of volcanic origin, similar to terrestrial basalt;
—fine-grained particles which have been subjected to severe temperature variations, radiation, meteorite bombardment; these therefore occurred on the surface or immediately below.

The first category looks surprisingly "fresh", like recently splintered basalt. The particles are angular in shape.

The greatest variations occur in the second category, where the most irregular shapes are found, while the surface of the grains contain vitreous and metallic globules, which the Soviet scientists called "cosmic globules".

The chief minerals found in the first category were: plagioclase, olivine, pyroxene and ilmenite, which also form the four main elements of solidified rock on earth.

The "vitreous" globules vary in colour: greenish, dark yellow cloudy white and transparent. They must have been formed in very high temperatures and probably emanate from the inside of craters.

As was the case with the Apollo-11 material, no unknown elements were encountered in the Russian samples, nor had any been expected.

The following is a table comparing the chemical composition

of the Luna-16 samples with that of the material brought back by the Apollo-11 crew:

	Basalt		'Dust'	
	%	%	%	%
	Luna-16	Apollo-11	Luna-16	Apollo-11
SiO_2 (silicon-oxide)	43·8	40	41·7	42
TiO_2 (titanium-oxide)	4·9	3·7	3·39	3·1
Al_2O_3 (aluminium-oxide)	13·65	11·2	15·32	14
FeO (iron-oxide)	19·35	21·3	16·8	17
MgO (magnesium-oxide)	7·05	11·7	8·73	12
CaO (calcium-oxide)	10·4	10·7	12·2	10
Na_2O (sodium-oxide)	0·33	0·45	0·37	0·40
K_2O (potassium-oxide)	0·15	0·065	0·10	0·18
MnO (manganese-oxide)	0·2	0·26	0·21	0·25
Cr_2O_3 (chromium-oxide)	0·28	0·55	0·31	0·41
ZrO_2 (zirconium-oxide)	0·04	0·023	0·015	0·09

It strikes one that in some cases there are considerable differences. The same applies to a comparison between Apollo-11 and -12 samples, in which there was, in particular, a great difference in the titanium oxide content. In the Apollo-11 samples this varied between 4·5 and 7 per cent! In all, the Russians identified 70 chemical elements in their lunar samples. The Americans traced 66.

Luna-15

Luna-16's success confirmed—if, indeed, this was necessary—the suspicion that at the time of the first American flight to the moon (Apollo-11), the Russians had also tried to lay their hands on lunar material, if possible before the Americans could do so. Their Luna-15 went into lunar orbit at the time when Armstrong, Aldrin and Collins took off from Cape Kennedy, so the Luna was about three days ahead. The unmanned probe was to make a soft landing on the moon after a number of orbits, to scoop up some samples of lunar soil and return to the socialist fatherland. This plan did not materialise: after 52 orbits the Luna was decelerated, but not sufficiently so. On July 21, the day Armstrong first set foot on the moon, the Russian capsule crashed at a speed of 480 km per hour.

Shortly before the flight of Luna-16 I had a talk with the space-walker Alexei Leonov in Moscow. I told him that

Vladimir Komarov had once told a reporter of *Paris Match* that the Russians would land a recoverable "lunar soil snatcher". Leonov did not at first want to confirm this. He tried to evade my questions with the remark: "Tsiolkovsky already had that idea."

"Yes," I replied, "but that was only a dream, not a plan."

Leonov: "We have already landed robots on the moon which have photographed the surface in detail. They also measured the soil resistance, which was a difficult operation. Sending a robot is therefore no longer a complicated problem."

"But I should have thought that you'd need samples as well as photographs," I said.

Leonov smiled mysteriously. "I'm all for it. We'll get them."

At that moment I was completely convinced that Luna-15 should have taken lunar samples, the more so as Professor Sedov had assured me that for the time being the Russians did not intend to send cosmonauts to the moon.

Lunokhod-1: a car on the moon

When, shortly after the successful Luna-16 experiment, American experts were asked what they thought of this Russian achievement, they usually replied that they certainly appreciated its value, but that the activities of this moon robot could not remotely be compared with that of astronauts of flesh and blood. Luna-16, said the Americans, took a 100 g lunar sample from the spot where it happened to have come down. A human being, on the other hand, can move and look round to see what seems to be most interesting to him.

Less than two months after the return of Luna-16 this argument was considerably counteracted by the landing of Luna-17, which deposited Lunokhod-1, the first (unmanned) moon-car. In the months that followed, this robot showed convincingly that it was capable of exploring a large area, even though it could not bring samples back to earth.

The Russian moon vehicle proved to be a stout piece of machinery. On November 17, 1970, it rolled from the Luna-17 descent stage* down ramps which had previously extended. Even after its three months' period of planned activity, it continued to travel around in the extreme west of the Sea of Rains,

*This was a standardised descent stage, similar to that of Luna-16.

an almost level lava plain, which forms the "right eye" of the moon's face. Lunokhod-1 became "champion moon explorer" in that it holds the long-term record. Clear indications that its activities were coming to an end were not received until the middle of September. When, on October 1, 1971—that is, nearly a year after the vehicle had been deposited on the moon —the sun's rays lit up the lunar mountains on the western bank of the Sea of Rains once more, interested parties on earth waited in suspense for news about the robot's condition. On October 10 Tass announced that efforts to re-establish contact with Lunokhod-1 had ceased six days previously. The vehicle had covered a distance of 10 540 metres. For ten months on end it had repeatedly withstood the intense cold of the lunar night (minus 150 degrees Celsius) and the heat of the lunar day (over 100 degrees).

During its first three months on the moon Lunokhod-1 covered over five kilometres. It would easily have been capable of covering a much greater distance in that time, but of course its mobility in itself was not the chief objective of the experiment. Its main purpose was to take measurements. The Lunokhod therefore moved a few metres at a time. When it stopped, its "sensors" went into action. Its chief sensor was the "Rifma", which is an abbreviation of the Russian equivalent of "x-ray spectrometer". This apparatus, a small box mounted between the front wheels, was equipped with radio-active cells which directed hard radiation into the soil. Chemical elements in the soil responded by emitting x-rays. From the intensity and the nature of the reflected radiation it was possible to deduce the chemical composition of the soil, although this method produced less detail than analysis in a laboratory on earth. In any case scientists could determine the aluminium, silicon, magnesium, potassium, calcium, titanium and iron content of the soil. All these elements proved to be present, usually in combination with oxygen. The Rifma, which had previously been calibrated on earth, is used in a less refined version in blast furnaces, where it rapidly analyses metals and slag.

The Lunokhod also had a "feeler" in the form of a penetrometer, which was plunged into the soil at different places. The resistance encountered provided information about the bearing strength of the lunar surface. The Lunokhod was further equipped with an instrument for measuring cosmic

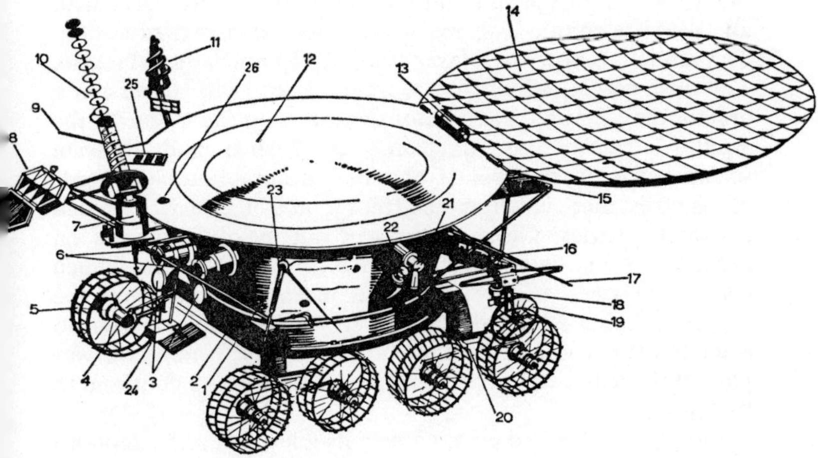

Fig. 20. Lunokhod-1.

The first lunar vehicle has the same dimensions as a mini-car: wheelbase 2216 mm; width of track 1600 mm; diameter of wheels 510 mm. The width of the instrument container is 2150 mm at its upper surface. Total weight: 757 kg.

1. Hermetically sealed instrument container.
2. Reinforced underside of container, which acts as chassis.
3. Protective cover for television cameras. Below these, the Rifma apparatus.
4. Electric motor (one in each wheel).
5. "Lunar wheel" with wire mesh surface (not visible).
6. Television cameras.
7. Aerial direction mechanism.
8. (French) laser-reflector.
9. Extensible antenna.
10. Directional aerial (sensitive in one direction only) for transmitting TV pictures.
11. Omni-directional aerial (sensitive in all directions).
12. Upper surface of container, which at the same time acts as a radiator for cooling the apparatus.
13. Hinge mechanism of "lid".
14. "Lid", containing solar cells to provide electricity.
15. Bracket for "lid".
16. Isotope source for heat energy.
17. Extensible antenna.
18. Penetrometer (apparatus for determining bearing strength of surface).
19. Trailing wheel for measuring both distance covered and slip differential as compared with drive wheels.
20. Pivoting bogey arm for each set of two wheels.
21. Telephoto-camera for taking vertical pictures, showing the lunar surface and the wheels.
22. Panoramic telephoto-camera.
23. Extensible antennae.
24. Rifma (X-ray spectrometer).
25. X-ray telescope.
26. Radiation meter.

radiation—from the sun and from the Milky Way—and with an x-ray telescope for registering cosmic x-ray radiation. The vehicle also carried a laser reflector of French manufacture. This instrument, which has a reflective capacity three times greater than that of the reflector deposited on the moon by the Apollo-11 crew, was monitored by French and Russian scientists for the purpose of obtaining accurate measurements of the distance between the moon and the earth, which constantly varies. Laser beams sent out by observatories on earth were reflected back to earth. From the time which elapsed between transmission and reception, and from the speed of light, which is known, it was a simple matter to calculate the distance. At the end of its 'trip' Lunokhod was parked in such a way that the laser reflector could continue to function.

Lunokhod also had eyes, namely its television and telephoto cameras. These eyes served, as it were, as extensions of the human eyes 400 000 km away, for the moon-car was driven by a team of five technicians—in the Centre of Remote Space Communications in the Crimea.

At first these men, like men on the moon, were rather tense, but gradually they acquired great expertise in working with the moon-car. Whereas on its first moon-day the lunar explorer was made to cover only 197 metres, the total distance covered by the end of the second moon-day had increased to 1719 metres. On this second "day" the Lunokhod went into second gear for the first time. The car's top speed was never published, but doubtless it was very low. It takes over one second for a signal from earth to reach the moon, and vice versa, and a lot can happen in that time. The Lunokhod did have an automatic brake, which functioned every time the vehicle reached too great an incline, for instance when it approached a very steep crater-wall. The first working day lasted until November 27; then the Lunokhod was parked. By means of batteries the temperature in the instrument container (in the west sometimes called "the bathtub") was maintained at around 15 degrees Celsius. The solar cell panel which throughout the lunar day had been directed towards the sun to transform sunlight into electric energy, was closed up during the night in order, as far as possible, to prevent heat loss.

On December 9, when the sun had once more risen over the landing area, the panel was again deployed towards the sun to

V. (*Above*) As Soyuz-4 and -5 orbited the earth after docking, two cosmonauts transferred from one spacecraft to the other.
(*Below*) Vladimir Shatalov in the cabin of the Soyuz-trainer.

VI. (*Above*) Soyuz: the first manoevrable Soviet spaceship.

(*Below*) Work compartment of the Soyuz. The hatch used for extra-vehicular activities is clearly visible.

recharge the batteries and the Lunokhod resumed its transmission of panoramic and other views of the moon's surface.

On December 10 the lunar car traversed a crater with a diameter of 16 metres and a depth of two metres. At one moment during its descent, the vehicle moved at an angle of 27 degrees. In the course of this second lunar day, the robot moon scout explored in all forty craters. In that period it obediently carried out no fewer than 4000 instructions. At 73 points it measured the soil resistance, which proved to vary with the distance from a crater. In general the soil is more powdery in structure on crater walls. By the end of the second period of activity, the Lunokhod was at a distance of 1370 metres from the Luna-17 descent stage.

Its second "night" on the moon lasted from December 23 to January 7. On January 8, when the sun stood four degrees above the horizon, the car went into action once more. "*Sdobrim utrom Lunokhod!*" (good morning Lunokhod), the Russian papers exulted. Controlled from earth, the robot cosmonaut went to work again without delay. Every twenty or thirty metres it stopped for a moment to measure the soil resistance and at interesting points it halted to determine the chemical composition of the soil. At the same time the other experiments continued: temperature, cosmic radiation, x-ray readings. . . .

Lunokhod remained in top-class condition. The pressure in its internal compartment, which was filled to the brim with electronic apparatus, was 755 millimetres, its temperature a comfortable 18 degrees C. (The apparatus could function at between 10 and 30 degrees C.)

On January 9, Lunokhod-1 spent half the morning in exploring an old crater with a diameter of 100 metres and a depth of 6 metres. Its earthly masters had soon learned to judge from their appearance which craters were (relatively) old and which more recent. When it had concluded its exploration the moon-car, as usual, transmitted panoramic telephotos to the large dish antennae in the Crimea. Whereas so far it had moved mainly in a south-easterly direction, it now changed course towards the north-east.

At that moment an extremely important navigational experiment began. The distant "crew" of the Lunokhod wanted to attempt to return the vehicle to its starting point—the descent stage of the Luna-17, nearly 1400 metres away. To do so, they not only made use of the inertial platform* in the Lunokhod, but also of the telephotographs and especially of the shadows of certain sections of the moon-car visible on these photographs. Since the position of the sun was known (two of the four telescopic cameras in fact took pictures of the sun and the earth in the moon's sky), Lunokhod's direction could be accurately established. By January 12 the vehicle had covered 2930 metres. Between January 13 and 15 it stopped: at that time the sun was practically at its zenith and consequently there were hardly any shadows. On January 17 radio contact was made with the lunar car for the 42nd time. It should be mentioned that radio contact was possibly only when the moon was above the horizon as seen from the Crimea. The Lunokhod covered another 254 metres, taking a large number of smaller craters in its stride (see fig. 20).

The navigator proved to have done his task well. During the 43rd "session", at 0.30 Moscow time on January 18, the descent stage appeared in the picture. One of the telescopic cameras took a beautiful photograph, showing Luna-17 as well as the earth. It had been clearly proved that the Soviets could direct their automatic cosmonaut wherever they wanted it to go. This would certainly be of very great importance in future, even more complicated, experiments.

Towards the end of the third lunar day, during which the vehicle covered 1936 metres, Lunokhod-1 again set off. This time it took a north-westerly direction to explore the area north of the landing site. From January 26 to February 7, the robot experienced its third lunar night and came through unscathed. At that time 3655 metres had passed beneath its eight wheels. On February 8, as it was traversing a landscape strewn with craters, it completed its fourth kilometre.

Because of the high altitude of the sun it stopped again from February 12 to 15. By then it had covered 4813 metres in all. The exact distance was measured with the aid of a ninth (trailing) wheel at the rear of the vehicle, which pressed against the lunar surface with a certain amount of force. The rim of this wheel was spiked, so the problem of slipping was reduced

*An inertial platform consists of three gyroscopes with axes at right angles to each other. A gyroscope is a rapidly revolving top, independently suspended, which constantly maintains the same position. The movements of the vehicle housing the gyroscope can be measured in relation to it.

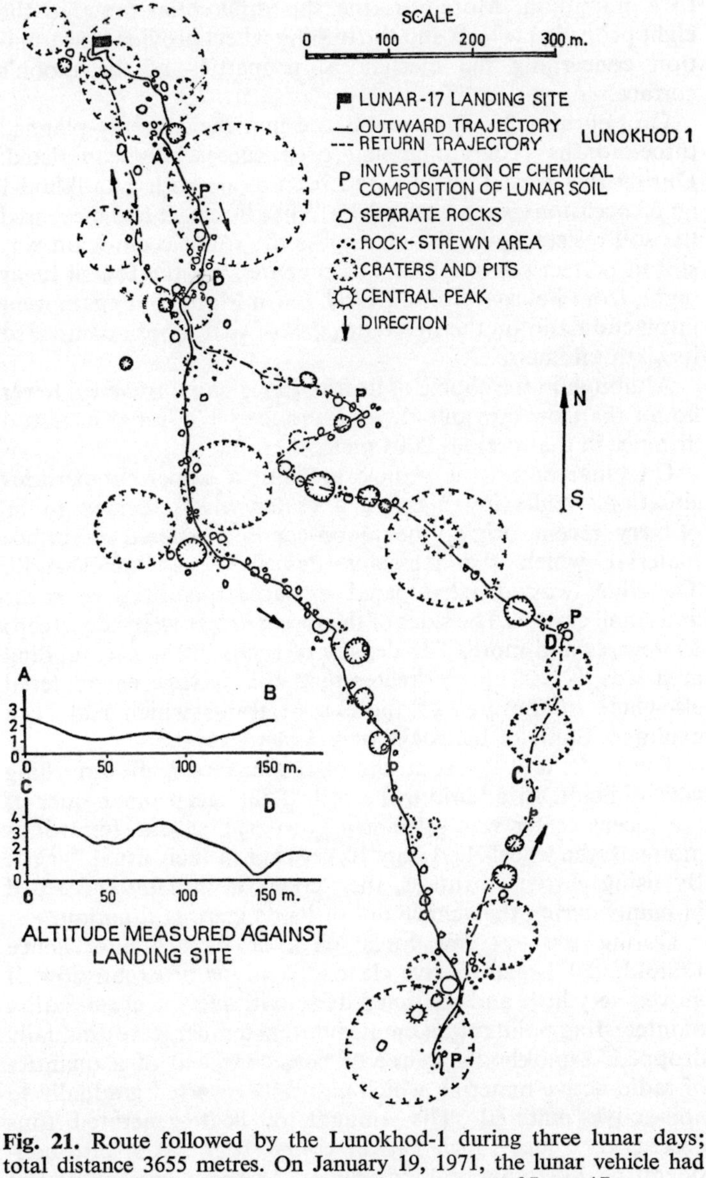

Fig. 21. Route followed by the Lunokhod-1 during three lunar days; total distance 3655 metres. On January 19, 1971, the lunar vehicle had returned to its point of departure: the descent stage of Luna-17.

to a minimum. Moreover, the slip differential between the eight propelled wheels and the trailing wheel provided information concerning the mechanical properties of the moon's surface.

On February 20, Tass announced that the carefully-planned three-months programme had been successfully completed. During this period contact had been made with Lunokhod-1 on 63 occasions, each lasting two to five hours. It had measured the soil resistance in 200 places. As the robot cosmonaut was still in perfect condition, it was prepared for the fourth lunar night, from February 20 to March 6. On March 8 it again went into action and on the thirteenth day of that month completed its sixth kilometre. . . .

Although in the course of its fifth lunar day it travelled fewer hours than on previous days, Lunokhod-1 covered a record distance in that period: 2004 metres.

On lunar day 6 the vehicle got into a somewhat perilous situation. While investigating a crater which seemed to be of very recent origin, the moon-car encountered a surface material which the Russians described as "quicksand". Travelling was also hampered by large quantities of rocks and small craters. The sides of the recent crater were very steep: 30 degrees and more. The density of rocks in the surrounding area was 80–100 times greater than the average encountered elsewhere in the area of the Sea of Rains which had been explored. Some of the rocks were 3 metres high!

The rocks and the contours of the surface made travelling very difficult. In addition, the soil of the steep inner sides of the recent crater was extremely powdery: at one hazardous moment, the wheels had only 10 per cent of their usual "grip". By using extreme caution, the "crew" eventually succeeded in manoeuvring the vehicle out of its dangerous situation.

During the last few lunar days of its active existence Lunokhod-1 began to give clear indications of exhaustion. It moved very little and restricted its activities to the examination of interesting points close by. Its internal temperature gradually dropped. Lunokhod's source of heat consisted of a quantity of radio-active material, which naturally reverted gradually to non-active material. The amount of heat generated thus decreased, a fact which had of course been taken into consideration by Lunokhod's designers. In the course of its last lunar day the vehicle covered approximately 100 metres.

Meanwhile the Russian scientists of the Academy of Sciences were fully occupied in digesting the enormous volume of information which Lunokhod had sent its masters up to that time. The vehicle had explored an area totalling no less than 80 000 square metres. Over 200 panoramas and nearly 20 000 photographs had been received in the Crimea. At more than 500 points Lunokhod-1 had determined the mechanical properties of the soil, while at 25 points it had analysed the chemical composition. According to Tass, the technical information received made it possible to perfect the design of succeeding vehicles.

Unmanned scouts are less versatile and selective than man. If one wants to know exactly what the Sahara is, one may have a number of ingenious instruments placed there, which send out a constant stream of information. But one would get a much better idea by taking a look oneself—and that preferably over a period of some time.

Nevertheless unmanned moon scouts undoubtedly have their advantages so long as man's opportunities of exploring the moon are limited. And they are cheap, which is an important consideration. Operations like those of Luna-16 and Lunokhod cost only a fraction of the amount spent on one manned flight to the moon. An unmanned lunar vehicle can go on moving as long as it has enough electricity and thermal energy provided there is no mechanical failure. An unmanned appliance does not get tired, it does not run out of oxygen, food or water. One single launching can therefore yield considerable results. An unmanned craft can moreover be sent to areas where astronauts cannot yet go.

On February 5, 1971, the American astronauts Alan Shepard and Edgar Mitchell landed on the moon in the lunar module of Apollo-14. For their second moon-walk the exploration of the Cone crater, 1200 metres away, had been planned. They just failed to reach the rim of the crater, because the medical men on earth considered it advisable to tell the tired men to go back. Naturally the Soviet press referred to this fact in order to prove that unmanned scouts are better for this kind of task. They did, of course, omit to say that on their next trip to the moon the Americans would take a lunar vehicle with them, which would carry them at 16 kilometres an hour.

Nevertheless, in spite of the latter, the scope for manned

flights to the moon, such as those in the recent American series, is limited both from a technical and from a financial point of view. After Apollo-17 no American spaceship—not even an unmanned one—will land on the moon for several years to come.

The Soviets, on the other hand, will certainly deposit even more highly perfected robots on the moon; apparatus which will exceed the combined achievements of Luna-16 and -17. It cannot be doubted that, once they have developed a complete space transport system, the Russians will send cosmonauts to the moon as well.*

Lunas-18, -19, -20, and -21

Lunas-16 and -17 had both come down in a relatively flat area, namely one of the lunar seas.

The first Russian attempt to land in mountainous terrain was that of Luna-18, which was launched on September 2, 1971.

The Russian scientists regarded a landing on one of the continents of the moon as very important, since obviously these areas are much older than the lunar seas.

Unfortunately the Luna-18 operation was not a success. On September 11, 1971, at 10.48 Moscow time, the Luna reached a spot in the mountains north of the Sea of Fertility, where Luna-16 had landed. Luna-18 crashed at 3 degrees 34 minutes northern latitude, 56 degrees 30 minutes eastern longitude. Probably it descended incorrectly in mountainous terrain.

Luna-19 was launched on the 28th of the same month. Expectations that this spacecraft was also intended to make a soft landing were belied in Tass's first communiqué, which alluded to "exploration of the moon from lunar orbit". Luna-19 therefore became a lunar sputnik. For many months the satellite relayed surprisingly sharp panoramic photographs, which helped scientists in their choice of further interesting places for future soft landings. Luna-16's success was continued by Luna-20, which made up for the failure of Luna-18. Luna-20 landed at 22.19 Moscow time on February 21, 1972, at 3 degrees 32 minutes northern latitude, 56 degrees 13 minutes eastern longitude, that is, almost at the same spot where no. 18 had had such an unfortunate ending. Luna-20

*See also "Space Travel: the Soviet Approach", p. 23.

proved to be practically identical to Luna-16, although in some respects it had been perfected. Luna-20 made a softer landing than any of its predecessors, because the main engine for the final braking manoeuvre was already fired at a height of 760 metres (in the case of Luna-16 this was done at 600 metres). On the basis of a number of panoramic pictures scientists were able to select a spot for the main operation of the new "lunar geologist", that is, drilling for lunar soil in this very ancient area of the moon: near the crater Apollonius C, 130 km north of the place where Luna-16 had landed. All operations succeeded perfectly. On February 23 the return stage lifted off from the moon with its soil samples and two days later the small blackened capsule reached the snow-covered steppes 40 km north of Dzhezkagan in Kazakhstan, bringing lunar samples which were probably a thousand million years older than those collected by Luna-16. On January 16, 1973 Luna 21 delivered Lunokhod-2 on the moon. This improved vehicle was active during 4 months, covering a distance of 37km.

One "first" after another

With the exception of a manned moon-landing, the Russians have, since 1959, been first with every important operation in the field of lunar exploration. In September of that year their Luna-2 was the first man-made object to reach the moon. Luna-3 was the first to take photographs of the far side of the moon, which until that time had never been seen.* Luna-9 made the first soft lunar landing; Luna-10 became the first lunar satellite; Zond-5 was the first space vehicle to return to earth after looping the moon; and Luna-16 was the first robot to bring a lunar sample back to earth, although it is arguable whether this may be called a first, since American astronauts had done this previously. Up to now the Russians have used four different types of spacecraft for their unmanned moon exploration. Lunas-1 to -3 weighed only a few hundred kilograms. They were incapable of correcting their course, unlike Lunas-4 to -14, some of which weighed up to nearly two tons. Numbers -4 to -8, according to official Soviet sources,

*The photographs taken by Zond-3, a capsule similar to the "Venusnik", were of better quality.

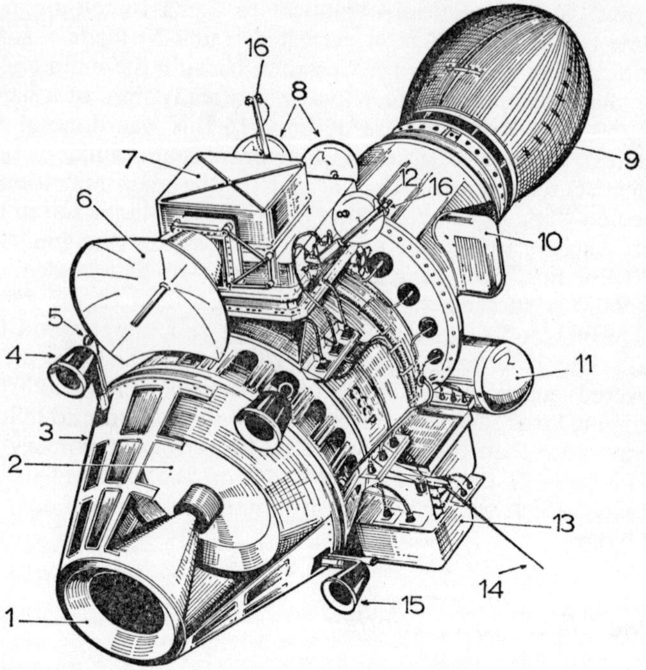

Fig. 22. Luna-9 (capsule with engine- and instrument section).

1. Liquid fuel engine for course corrections.
2. Fuel tanks, etc.
3. Engine cover.
4. Vernier engine for attitude control.
5. Attitude control jet.
6. Antenna for radio-altimeter.
7. Radio-altimeter with accessory electronic equipment.
8. Spherical gas tanks for attitude control system.
9. Instrument capsule (with petal antennae in closed position).
10. On-board computer and control apparatus.
11. Hermetically sealed instrument container.
12. Gas reservoir.
13. Stellar orientation equipment.
14. Antenna.
15. Vernier engine.
16. Micro-engines for attitude control.

were used for elaborating the soft landing technique eventually carried out with Luna-9. There are strong indications that Luna-9's five predecessors were all intended to have made a soft landing. By American standards Luna-9 did not, in fact, make a soft landing, but a semi-soft one. Near the moon's

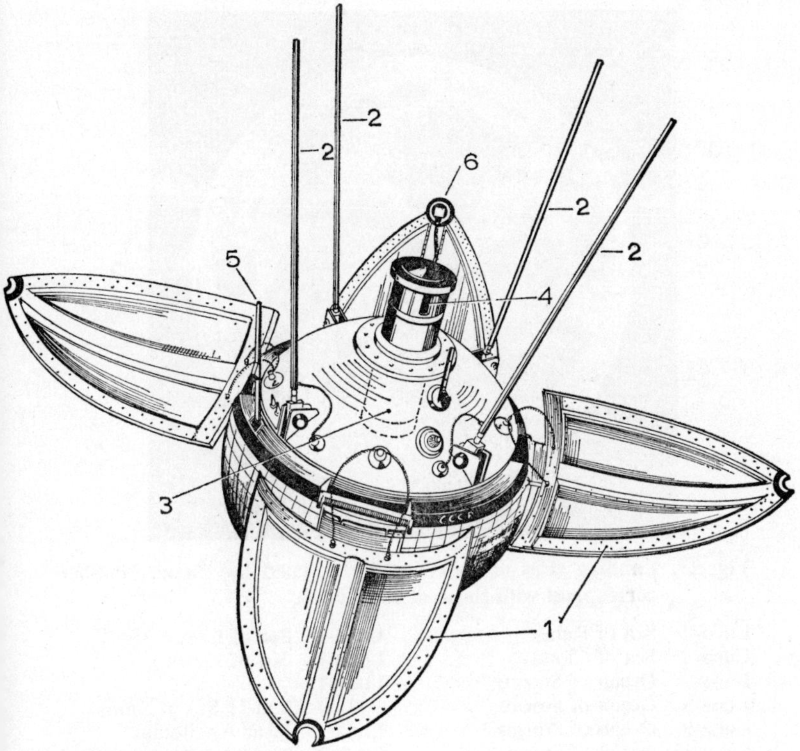

Fig. 23. Luna-9 capsule.

1. Petal antennae.
2. Retractable rod antennae.
3. Instrument container.
4. Television camera.
5. Rod which has two mirror surfaces (acts as "rearview mirror" for the television camera).
6. Disconnection mechanism for petal antennae.

surface the instrument capsule was separated from the retro-rocket section, after which the capsule, which was equipped with a shock-absorber, fell onto the moon. Luna-9 made the first photographs of the lunar surface showing 1 to 2 millimetre particles. The most important result of the experiment was the fact that the capsule did not, as some pessimistic scientists had predicted, sink down into a layer of dust. This proved that a human being would be able to set foot on the moon. Luna-13, a direct descendant of No. 9, moreover measured

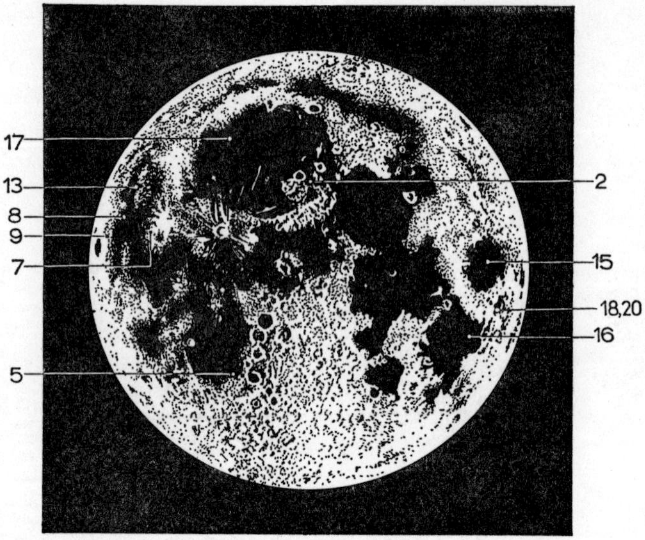

Fig. 24. Landing sites of Lunas which reached the moon. Numbers correspond with those of the Lunas.

Luna-2	Sea of Rains
Luna-5	Sea of Clouds
Luna-7	Ocean of Storms
Luna-8	Ocean of Storms
Luna-9	Ocean of Storms
Luna-13	Ocean of Storms
Luna-15	Sea of Crises
Luna-16	Sea of Fertility
Luna-17 (Lunokhod-1)	Sea of Rains
Luna-18	Near Apollonius
Luna-20	Near Apollonius

the bearing strength of the lunar surface by driving a pin into it by means of an explosive charge. The basic elements of Luna-9 and -13 were used in Lunas-10, -11 and -12, which all went into orbit round the moon, from where they measured, among other things, its gravity field and its magnetic field and in addition provided clear panoramic photographs.

Lunas-15, -16 and -17, all much heavier than their predecessors, belong to the third category.

Zond-4 and subsequent capsules may be said to belong to the "very heavy" type. They bear no relation to the Lunas, which are purely unmanned probes. This does not apply to the Zond, which could be manned. The return capsule of these spacecraft is practically identical to that of the Soyuz (Zond-5 and 7, carried, among other things, tortoises). The Russians would therefore be capable of making a manned flight round the moon in a Zond. (Note: *round* the moon, for certainly none of the vehicles of this type launched so far could have landed on the moon and returned.) In fact no Zond has yet been injected into lunar orbit, although from a scientific point of view this would have been of the greatest importance. Naturally far more information may be gathered during, say, ten orbits round the moon (as in the case of Apollo-8) than by describing one loop around it. The former, however, does mean that the explorer vehicle must be equipped with a powerful engine, capable of decelerating the satellite (which weighs several tons) to bring it into the sphere of influence of the moon, and subsequently to achieve (lunar) escape velocity to enable the spacecraft to return to earth.

For the Zond flights, two different recovery procedures were applied, one making use of the capsule's aerodynamic properties, the other without. In the latter case (used for Zond-5), the capsule dives steeply into the atmosphere and lands. Such a ballistic re-entry subjects the capsule to maximal deceleration, which no cosmonaut could survive. On an aerodynamic re-entry, deceleration is gradual, so that the

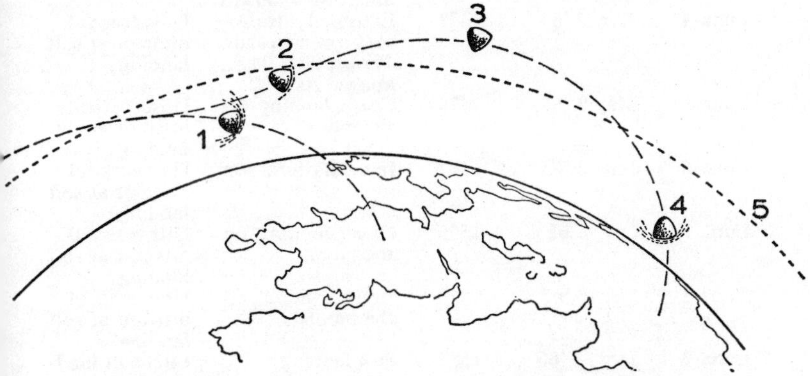

Fig. 25. Ballistic and aero-dynamic return of a Zond capsule.

1. Ballistic return with maximum deceleration.
2. Aerodynamic return, making use of the capsule's "lift" capacity.
3. Highest point of trajectory, outside the atmosphere.
4. Final braking stage with minimal deceleration forces.
5. Upper layers of the earth's atmosphere.

cosmonaut is subjected to no more than four to seven times his own weight. The aerodynamic deceleration technique requires capsules of suitable shape, such as that of the Zond (Soyuz) capsules. By varying the angle between flight path and heat-shield, the upward force can be controlled. This procedure was followed in the Zond-6 and subsequent operations.

Survey Luna Flights

Name	*Launched*	*Weight (kg)*	*Flight*	*Details*
Luna-1	Jan. 2 '59	361·3	In orbit round sun	Unsuccessful attempt to hit moon?
Luna-2	Sept. 12 '59	390·2	Crash landing on the moon	First object on the moon
Luna-3	Oct. 4 '59	278·5	Extended trajectory round earth. Perigee 40 300 km, apogee 476 500 km	First photographs of far side of moon
Luna-4	Apr. 2 '63	1422	Extended trajectory round earth. Perigee 90 000 km, apogee 700 000	Unsuccessful attempt at soft landing?
Luna-5	May 9 '65	1476	Crash landing on the moon	Unsuccessful attempt at soft landing?
Luna-6	June 8 '65	1442	In orbit round sun	Unsuccessful attempt at soft landing?
Luna-7	Oct. 4 '65	1506	Crash landing on the moon	Unsuccessful attempt at soft landing?
Luna-8	Dec. 3 '65	1552	Crash landing on the moon	Unsuccessful attempt at soft landing?
Luna-9	Jan. 31 '66	1583	Soft landing	First soft landing. Photographs of moon surface
Luna-10	March 31 '66	1600	In orbit round the moon	Information on gravity- and magnetic-field of the moon. Ditto on gamma radiation

Luna-11	Aug. 24 '66	1645	In orbit round the moon	Information on gravity- and magnetic-field of the moon. Ditto on gamma radiation
Luna-12	Oct. 22 '66	1600?	In orbit round the moon	Panoramic photographs of lunar surface. Information on gravity-field and radiation
Luna-13	Dec. 21 '66	1600?	Soft moon-landing	Soft landing. Photographs of lunar surface and measurement of mechanical properties of the surface
Luna-14	Apr. 7 '68	1600?	In orbit round the moon	Measurements of lunar gravity-field and radiation
Luna-15	July 13 '69	1880?	First in lunar orbit, followed by crash landing	Unsuccessful attempt at soft landing and at taking lunar soil for examination on earth
Luna-16	Sept. 12 '70	1880	First in lunar orbit, followed by soft landing on Sept. 20. Back on earth Sept. 24	Soft landing. Soil sample taken and delivered to earth
Luna-17	Nov. 10 '70	1900?	Soft moon landing and depositing Lunokhod-1 (weight 756 kg)	Successful. Moon-car was active for ten months
Luna-18	Sept. 2 '71	1880?	In lunar orbit on Sept. 7. Attempt to make soft landing on Sept. 11	Soft landing failed in mountainous territory
Luna-19	Sept. 28 '71	1880?	In lunar orbit on Oct. 3	Photographs of lunar surface from orbit
Luna-20	Feb. 14 '72	1880?	First in lunar orbit; followed by soft landing on Feb. 21. Returned to earth on Feb. 25	Soft landing in mountainous territory. Lunar soil delivered to earth

Luna-21	Jan. 8 '73	2000?	First in lunar orbit; soft landing on Jan. 16	Successful. Second moon-car (840 kg)

Survey Zond Flights

Name	*Launched*	*Flight*	*Details*
Zond-1	Apr. 2 '64	In orbit round sun	Relayed information on interplanetary space. Unsuccessful shot at Venus?
Zond-2	Nov. 30 '64	In solar orbit	Shot at Mars. Premature loss of contact
Zond-3	July 18 '65	In solar orbit	Took photographs of far side of the moon
Zond-4	March 2 '68	Extended orbit round earth	Flight round imaginary moon? Landed on March 9 in Indian Ocean. Not recovered
Zond-5	Sept. 15 '68	Round back of moon	Photographs of earth and moon. Landed in Indian Ocean on Sept. 21
Zond-6	Nov. 10 '68	Round back of moon	Photographs of earth and moon. Landed in Kazakhstan on Nov. 17
Zond-7	Aug. 8 '69	Round back of moon	Colour photographs of earth and moon. Landed in Kazakhstan on Aug. 14
Zond-8	Oct. 20 '70	Round back of moon	Colour photographs of earth and moon. Landed in Indian Ocean on Oct. 27

AN EYE ON DISTANT WORLDS

Venera-7: success – The difficult survey of Venus – The technical dilemma – Venera-8, Mars-2 and 3

Centre for Remote Space Communications (Crimea). It is nearly eight o'clock Moscow time on December 15, 1970.

The high-pitched signal, which until a short time ago sounded from the loudspeakers, has stopped. Sixty million kilometres away a spherical capsule has separated from the Venera-7. As a result of this manoeuvre, the parabolic antenna of the parent module has lost touch with the distant earth, so that the powerful dish antennae in the Crimea can no longer receive a signal. Now one has to wait for the transmitter of the capsule which is racing towards the atmosphere of Venus at a speed of 11 km per second. The parent module has ceased to play a part. The capsule should now have reached the upper layers of the dense Venusian atmosphere. Deceleration should be maximal (400 times normal earth gravity), the temperature 10 000 degrees Celsius. That is the temperature on the surface of the sphere, which is covered in a heat-resisting material. The temperature inside the capsule is normal room temperature. Shortly before entry into the atmosphere of Venus, the internal temperature of the Venusnik has been lowered to minus eight degrees Celsius.

As a result of atmospheric resistance, the speed should now have decreased to 250 metres per second. At this moment the parachute should be deployed, which is to lower the capsule slowly to the surface of the planet. Where is the signal? Are both antennae deployed? "There's the signal!" someone shouts. The time is exactly 8 hrs. 4 mins. 21 secs.

Will Venera-7 reach the planet in good shape? Will the capsule and its contents be able to withstand the barbaric conditions there: around 500 degrees Celsius and an atmospheric pressure of over 100? Three previous Veneras failed after a successful initial descent. This new capsule is heavier and its insulation has been improved. The technicians and journalists, gathered round the receiver, are waiting in suspense.

While the capsule swings down through the atmosphere under its scarlet-edged parachute, faster than its predecessors, it transmits details of the rapidly increasing temperature and pressure and about the composition of the atmosphere of the planet. The report takes exactly 35 minutes, then the transmitters are silent. The surface must now have been reached. The machines on which the signals are recorded continue to revolve. But no signal comes. Will Venera-7 have the same fate as its predecessors? It is beginning to look like it.

Tass's first reports, issued on December 18, do not mention whether any signals have been received from the surface of the planet. It appears obvious what conclusions should be drawn.

But over a month later, on January 27, it is announced that Venera-7 most certainly showed signs of life after reaching the planet—for no less than 23 minutes in fact.

However, the signal's strength was a hundred times weaker than it had been during the capsule's descent. A computer had to be used to amplify the signals so that they could be interpreted. Hence the Russians' uncertainty immediately after the landing. Probably the probe had landed in such a way that the antenna was incorrectly aligned, so that only a very small part of its transmitting power—slight in itself—could reach the earth.

But in the long run this was of little importance. Venera-7 was the first to prove with certainty that the temperature on the surface of Venus was no less than 475 degrees Celsius (plus or minus twenty degrees), and the atmospheric pressure ninety (plus or minus 15). Thus, for the first time, information was received directly from the surface of another planet.

Venus: a planet difficult to approach

Ever since, on February 12, 1961, the Russians launched their first Venusnik, they have struggled ceaselessly to discover the secrets of Venus, the mysterious morning and evening star. It must be admitted that in the end they were reasonably successful. It is mainly owing to the Russian Venusniks that we know a great deal more about the atmosphere of our neighbour planet than we did some years ago. Unlike its

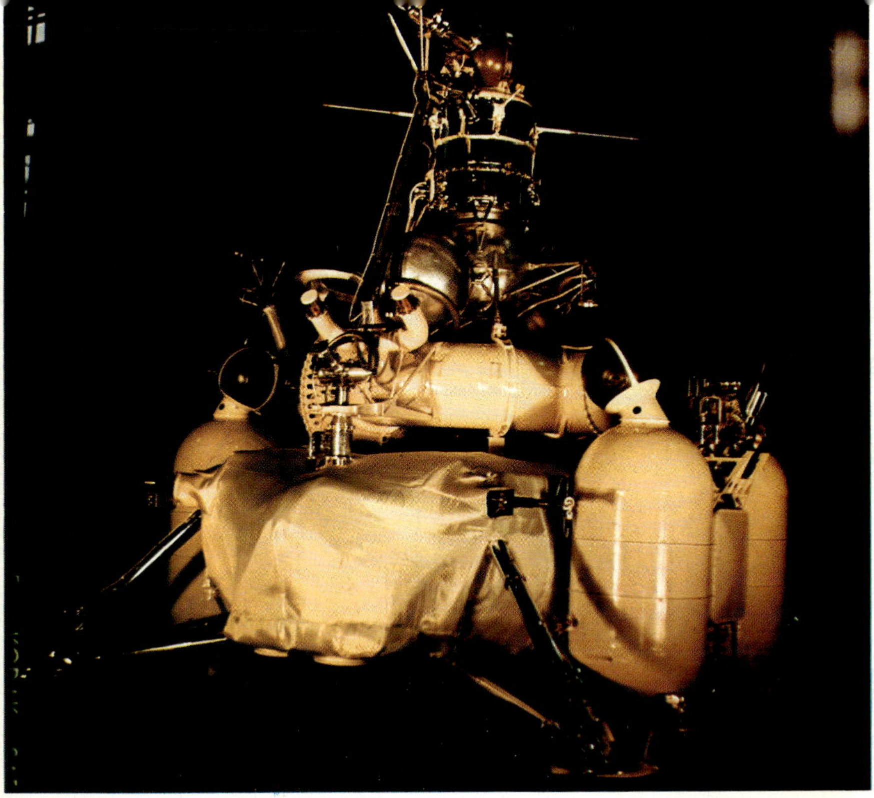

VII. (*Above*) Luna-16, the first unmanned spacecraft to bring lunar material to earth, ready to be mounted on the carrier rocket.

(*Below*) Prototype of Lunokhod-1 (as yet without instrument container), on the "Lunodrome", an artificial lunar landscape laid out in the factory where the vehicle was constructed.

mythological namesake, the planet keeps itself veiled in several layers of dense cloud. It cannot be denied that the Soviet Venera programme succeeded in lifting several of the veils, although no one can as yet say what the surface of the planet looks like. The last veil remains to be lifted. However, the Russian scientists can hardly be blamed for that; rather it is due to the fact that Venus makes the task exceptionally difficult. . . .

Venera-1, launched in February 1961, was the first shot at another planet. The intention was that after its three months' voyage, the capsule was to pass near to Venus. Due to loss of radio contact on February 27, the plan did not succeed. Although the Russians did everything in their power to restore contact (the Russian woman astronomer Alla Masevich went to Jodrell Bank with an assistant in an attempt to persuade the silent Venusnik to speak up again with the aid of the great radio telescope), they were unable to do so. The first Venera must have missed its objective by about 100 000 kilometres.

The Soviets then shifted their attention temporarily to Mars, but with little more success. Their Mars-1, launched on November 1, 1962, maintained contact with the earth up to a distance of 106 million km—in itself no mean achievement—but then fell silent.

Venera-2, however, maintained contact until, on February 27, 1966, it passed the planet at a distance of exactly 24 000 kilometres. At that critical moment the transmitter failed, so that the long awaited photographs were not received. Venera-3, launched four days after No. 2, on November 16, 1965, did not perform much better. Although the capsule of this probe reached the surface of the inhospitable planet, its transmitter failed on reaching the atmosphere, so that there can be little doubt that the instrument container, decorated with the Russian emblem, crashed at terrific speed.

VIII. *Left* A vision of the future, based on Russian information: a Soyuz "calls at" a large permanent space station. The station slowly revolves on its axis in order to produce artificial gravity in the cylindrical compartments at each end of the "spoke".

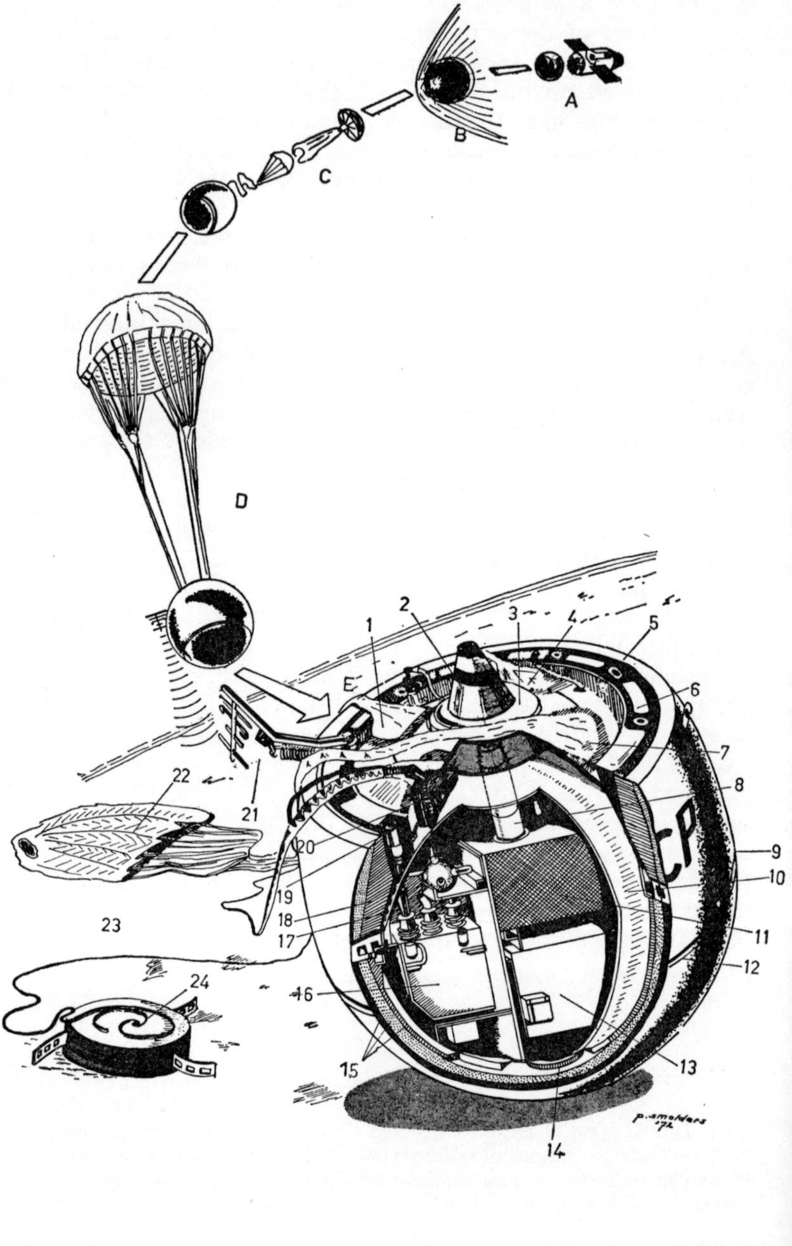

A
B
C
D
E
1
2
3
4
5
6
7
8
9
10
11
12
13
14
15
16
17
18
19
20
21
22
23
24
CP
p. smolders '72

Improved insulation

A greatly improved Venusnik departed on June 12, 1966. This Venera-4 fared much better than its predecessors. For the first time a capsule made a soft landing on another planet, in itself a magnificent "first". During its descent it transmitted information to the expectant Russians for no less than an hour and a half. At first the scientists thought that the capsule had fallen silent on reaching the surface, but later they reached the conclusion that Venus-4 had ceased transmission at a height of about 26 kilometres. At that height the temperature was 280 degrees Celsius and the atmospheric pressure 18. Measurement readings further showed that the atmosphere of Venus consists almost entirely of carbon dioxide, with a small trace of nitrogen, even less oxygen and hardly any water vapour.

Fig. 26. Venera 8.

Key to letters and numbers:

A. Heatproof capsule separates from command module.
B. Capsule becomes white-hot on entering into the dense atmosphere.
C. Cover is rejected. Drag parachute deploys main parachute.
D. Main parachute (capable of withstanding a temperature of 530 degrees Celsius) ensures soft landing. Radar antenna measures remaining distance from surface of Venus.
E. Capsule on Venus with a section of the casing cut away to show interior.

1. First parachute holder.
2. Antenna for communication with earth.
3. Gas analysator. Measures the composition of the atmosphere of Venus.
4. Second parachute holder.
5. External diameter of capsule at this point is 80 centimetres.
6. Collar for securing the cover.
7. Third parachute holder.
8. Ventilator.
9. Heat- and pressure-resistant covering.
10. Pressure chamber containing apparatus.
11. Electronic apparatus.
12. Ring-support to facilitate attachment to command module.
13. Radio transmitter.
14. Cushioning for pressure chamber.
15. Several layers of insulating material.
16. Adaptor for sending measurements to earth.
17. Radiator for cooling pressure chamber containing apparatus.
18. Internal insulation.
19. Explosive bolt for jettisoning hatch cover.
20. Fourth parachute holder.
21. Antenna of radio-altimeter.
22. Parachute.
23. Surface of Venus.
24. Second transmission antenna. Is jettisoned on landing.

This information indicated that it was essential to improve the insulation of the sensitive instruments, and if possible to accelerate the capsule's descent by making its parachute smaller. In this way the chances to penetrate further into the atmosphere and come closer to the planet's surface would improve considerably.

Veneras-5 and -6 in fact, launched on January 5 and 10, 1969 respectively, came a little closer to their objective. Their transmitters functioned up to a height of approximately twenty kilometres, where an atmospheric pressure of 27 and a temperature of 320 degrees Celsius were registered. By extrapolation of these figures the Russians concluded that on the surface the temperature must be around 500 degrees and the atmospheric pressure as high as 100. American estimates, based on the results obtained by the Mariner which flew past the planet, were in the same region.

This time the Soviets also obtained accurate details about the atmosphere of Venus: 97 per cent carbon dioxide, 2 per cent nitrogen, at most 0·1 per cent oxygen and around 0·05 per cent water.

"The chief problem facing us up till now," Professor Sedov told me before the start of Venera-7 during a conversation in his room in the Lomonosov University in Moscow, "was that we did not know enough about the most important characteristics of the Venus atmosphere. The recent flights have been of immense value for our further study of Venus. We have discovered that both temperature and pressure on the surface are enormous."

His statement led me to the conclusion that subsequent Veneras would doubtless be further improved, and in fact Venera-7 was an improved version of numbers 4 and 5, while its weight was increased by 50 kg. The greatest problem was still to prevent the delicate measuring apparatus and transmission equipment being "baked" before they reached the surface. As the chief designer of the Venusniks declared, it was the combination of high temperature and high pressure which made the technical problem such a challenging one. "The heat shield of the Venera has a double function, namely, to give out as much heat as possible through evaporation, so that the sphere remains intact; and to insulate the interior against the high temperatures which the capsule encounters in the denser layers of the atmosphere. But non-conducting materials are porous, and under the enormous pressure encountered in the lower levels of thc atmosphere, atmospheric gases penetrate porous material, heating the instruments and making them useless."

Thanks to the fact that before the descent the internal temperature of the container had been artificially lowered to minus eight degrees Celsius, *and* thanks to improved insulation and an even more rapid descent, Venera-7 eventually reached its objective in good condition. March 27, 1972 saw the departure of Venera-8.

Venera-8

Judging by the quantity of radio-active elements in the soil of Venus it would appear that the surface of the planet mostly resembles granite.

This was one of the most important conclusions of the investigations on the planet carried out by Venera-8. This probe was launched on March 27, 1972 and on July 22 made a soft landing, this time on the daylight side of the planet. For a period of 50 minutes this improved successor to Venera-7 transmitted information to earth. To make doubly sure, Venera-8 was equipped with a second transmitter antenna, which was ejected on reaching the surface of the planet and was linked to the capsule by a cable.

The new Venusnik reported that the surface of Venus is by no means hard. The density of the surface layer is only just over 1·5 grams per cubic centimetre. The comparative density of sandy soil on earth is 2·6 grams per cubic centimetre.

It appeared that on Venus, which revolves on its axis once every 243 days, day and night temperatures are practically the same. Venera-8, the first capsule to land on the illuminated side of the planet, recorded a temperature of 470 degrees Celsius.

One of the main reasons which decided the Russians to land their capsule on the daylight side on this occasion, was to find an answer to the question whether sunlight can reach the surface of the planet in view of the dense permanent cloud layer. Venera-8 proved that a hypothetical inhabitant of Venus would at any rate note a difference between day and night.

Venera-7's measurements of the planet's atmosphere were confirmed by Venera-8. There were moreover indications that there is a constant strong wind from the dark to illuminated side, blowing in the direction of the planet's revolution. These air movements occur chiefly at heights of some dozens of kilometres, so that they are hardly noticeable on the surface. There is no doubt that air currents have a greatly levelling effect on the temperature of the planet.

Mars-2 and 3

December 2, 1971, 16 hrs. 49 mins. 5 secs. Moscow time: the earth's first envoy to Mars lands in the haze of dust which conceals the crater-filled landscape of the bone-dry Mare Sirenum. The retro-rocket has done its work and a miniature rocket has cut the connection between the gigantic parachute and the descent module. Protective shields open like the petals of a flower and slow down the drop. The landing capsule of Mars-3 has reached the Red Planet.

Ninety seconds later the in-board computer gives the signal for action. At exactly 16 hours, 50 minutes and 35 seconds a first signal is relayed by the panoramic television camera on top of the module and recorded by miniature tape-recorders in the Mars-3 parent module, racing round the inhospitable planet at a great height. Exactly twenty seconds later communication is abruptly broken off. The Mars-3 capsule is silent. . . . Russian scientists believe that the failure of the landing capsule was most probably due to the violent dust storms raging in the area at the time of the landing. To substantiate this theory, photographs of the landing area made by the American Mariner-9, are sent to Moscow in the diplomatic bag.

This latest Russian Mars adventure had begun on May 19, 1971, when a "powerful carrier rocket" (probably an RNP) sent Mars-2, which weighed no less than 4650 kg, on its way to the Red Planet. The heavy weight indicated that this was no ordinary mission. The Americans' Mariner-9, which was already on its way to Mars, weighed only a ton, and even their Viking, which is intended to make a soft landing on Mars in 1976, will weigh only 3·5 tons. Part of this Viking is to orbit Mars, while another section will land on the planet.

Understandably it was thought that the Mars-2 and -3 capsules were also equipped for a soft landing.

In the course of their voyage to the planet, which took more than six months, the two "Marsniks" carried out all sorts of measurements in space. Like Lunokhod-1, one of the spaceships carried French experimental equipment, called "Stereo". This apparatus, mounted on one of the solar panels of Mars-3, recorded solar radio waves.

Mars-2 was the first to arrive in the region of Mars. On November 27 surmises were confirmed: a capsule separated from the parent ship and, by means of its own engine, corrected its course in such a way that it would make some sort of landing. This separation took place before Mars-2 went into orbit round the planet. After a solo-flight of more than $4\frac{1}{2}$ hours the capsule made a crash landing. Tass however announced that the Soviet emblem had been put on the planet Mars for the first time.

Meanwhile the Mars-2 parent module had gone into orbit round the planet, after having fired a retro-rocket. Its minimum distance from Mars was 1380 km, its maximum 25 000 km. Each 18 hours the Mars-2 module completed an orbit in this trajectory.

Five days later Mars-3 arrived and history repeated itself: on approaching Mars, a capsule separated from the parent module to land $4\frac{1}{2}$ hours later. This time, however, it was a soft landing. This raised the supposition that Mars-2 had also been meant to make a soft landing, all the more so as, according to the Soviet Academy of Sciences, Mars-2 and -3 were of similar construction and weight. It was, moreover, remarkable that only photographs of Mars-3 were released, none of Mars-2.

For their first landing on Mars the Russians, in principle, had the choice between two methods for slowing down their capsules: by using retro-rockets only, *or* by means of heat shield and parachute. The first technique was used for landing Luna-16 and its successors on the moon. However, the gravitational force of Mars is much greater than that of the moon, so that the comparative weight of the landing capsule would have required far more fuel for a soft landing on Mars. In its turn the extra fuel would have required a greater thrust from the carrier rocket—so much greater, in fact, that no existing rocket would have been capable of it.

The Russians therefore chose the combination of heat shield

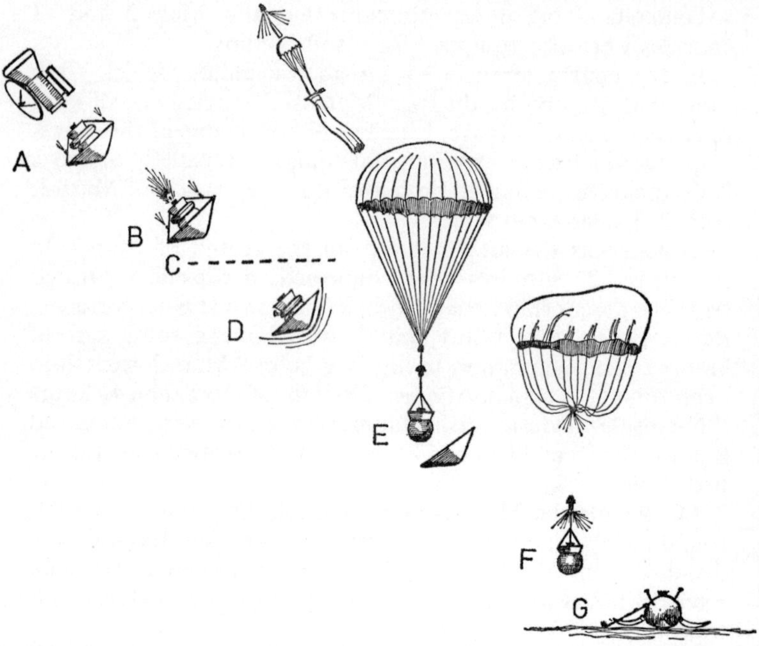

Fig. 27. Soft landing of Mars-3.

A. Separation of Mars-3 orbiter and landing capsule (see also detailed drawing, p. 235).
B. Solid fuel rocket engine propels the capsule, steered by small rockets, towards surface of Mars. Speed 6 kilometres per second.
C. Upper layers of the thin atmosphere of Mars (pressure approx. 7 mm mercury).
D. Aerodynamic deceleration through atmospheric friction.
E. Descent by parachute, speed having been decreased to 1200 kilometres per hour.
F. Ignition of retrorocket for soft landing. Parachute jettisoned.
G. Capsule with panoramic TV camera on top after landing on surface of Mars.

and parachute, which on Mars, with its thin atmosphere, is just feasible, although it cannot, of course, be used on the moon which has no atmosphere at all.

As a result of friction in the atmosphere, the conical heat shield slowed down the capsule, while at the same time protecting the instruments against overheating.

Although the soft landing on Mars provided little that was new from a scientific point of view, the parent capsules Mars-2 and -3 proved to be very fruitful in this respect. With their aid scientists were able to determine among other things, that:

> Mars' atmosphere consists chiefly of carbon dioxide, but its upper layers contain atomic oxygen and hydrogen. There are places on Mars where the temperature is up to 25 degrees higher than that of the surrounding area. The cause of this phenomenon is not yet known, but the possibility that there are active volcanoes is not excluded.
>
> The temperature on Mars at the terminator (the boundary between the illuminated and the dark portions of the planet) was minus 80 to 90 degrees Celsius. The highest temperature recorded was +15 degrees Celsius.
>
> Atmospheric pressure on Mars varies, according to position, between 3 and 8 millibars (on earth it is on average about 1000 millibars). On the basis of these figures the height of the terrain could be calculated.

Mars-2 and -3 also took photographs of the planet. Both capsules were equipped with a wide-angle and a normal television camera, which made (cartographic) panoramic pictures and photographed interesting smaller areas of Mars. The photographic activities in particular were at first seriously hampered by the storms which raged on the planet towards the end of 1971. As a result of an agreement between NASA and the Soviet Academy of Sciences, information was continually exchanged via a special "hot line" concerning the results obtained by Mars-2 and -3 and the American Mariner-9. After the storms had abated, the latter took thousands of very sharp photographs of Mars in addition to its other activities.

Nevertheless all this information together could still not settle the ancient question of whether there is (primitive) life on Mars or not. If the Mars-3 landing capsule had continued to function, it might have supplied the answer. In the event, video-signals were received for only 20 seconds. By domestic standards 20 seconds is sufficient for a T.V. advertisement, but when distances of millions of kilometres have to be overcome it is quite a different matter. The picture is transmitted dot by dot, line by line, in the form of figures; these are then decoded by computers on earth and are finally turned into a television picture. The Mars-3 picture received on earth was no more than a very narrow strip of a photograph.

It is possible that the Russians will land a capsule on Mars which will continue to function, before the Americans do so

Fig. 28. Detail of Mars-3.

1. Thrust impulse of main engine used for course correction and braking, to put orbiter in trajectory round Mars.
2. Container holding scientific apparatus for investigations of Mars and of the atmosphere surrounding the planet.

3, 4, 5. Star seekers of autonomous navigation system.

6. Solar cell panel.
7. Gas jet for attitude control.
8. Magnetometer.
9. Antenna for French stereo experiment.
10. Directional antenna for communication with earth.
11. Parabolic reflector of directional antenna.
12. Omni-directional antenna for communication with earth.
13. As above.
14. Gas tank for attitude control jet.
15. Antenna as 12 and 13.
16. Liquid fuel tank for main engine. This forms the central element of the structure.
17. Optical-electronic direction finder of autonomous navigation system.
18. Compressed gas tank (as 14).
19. Excess heat radiator.
20. Tube holding cooling liquid for radiator.
21. Second solar cell panel.
22. Attitude control gas jets.
23. Solid fuel rocket engine. Propels capsule towards planet (parent capsule continues at a tangent to orbital trajectory).
24. Antenna for communication with orbiter.
25. Fuel chamber for engine
26. Attitude control gas jet
27. Container holding parachute.
28. Radiator.
29. Surface of Mars (long distance view).
30. Conical heatshield for aerodynamic deceleration in atmosphere.
31. Band connecting Mars-capsule to parachute container.
32. Mars-capsule with panoramic TV camera and scientific apparatus.
33. Attitude control gas jets.
34. Strut between heatshield and rest of structure.
35. Small rocket which deploys parachute.

with their Viking (1976).* There are also plans to explore Mars with an unmanned mobile scout. Shortly after the Lunokhod-1 had landed on the moon, Professor Petrov said that one day there might be "planetokhods" as well. Subsequently one of the chief constructors of the Lunokhod, Vladimir Ivanovich, enlarged on this. "In my opinion," he said, "the Lunokhod opens up real opportunities for future planetary exploration." Asked to what extent the experience obtained with the Lunokhod would be of value in the construction of, for instance, a Marsokhod, he replied: "I think that the Marsokhod will differ essentially from the Lunokhod in its present form. In the

*At the time of publication Mars 4, 5, 6, 7 are on their way to the Red Planet. This series includes capsules which are to make soft landings by February-March 1974.

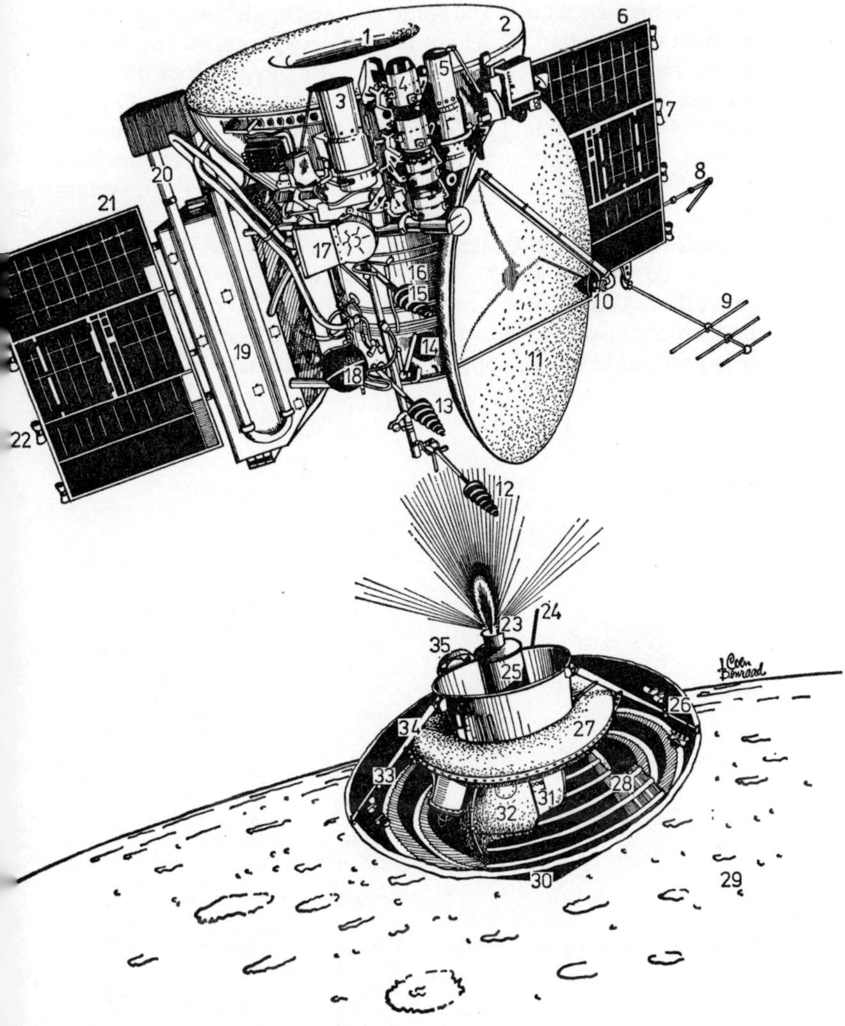

case of the moon a system exists for two-way communication, in which man plays the main role. Because of the immense distance involved, this will be difficult to achieve in the case of Mars. Consequently the Marsokhod will have to be far more self-sufficient. As far as direction control is concerned it looks as if we may encounter the same problems, although Mars has its own characteristics. There is an atmosphere, so the danger of dust storms cannot be excluded. This danger does not exist on the moon. On the other hand, scientists believe that the surface of Mars is less rugged and has less pronounced craters. In any case it is to be expected that our experiences with the Luna-17 experiment will be of value in future Mars probes."

According to the constructor it was unlikely that vehicles would be landed on Venus. "In view of conditions which prevail on Venus—a temperature of nearly 500 degrees Celsius and an atmospheric pressure of around 100 atms.—I believe it is preferable to use apparatus of the bathyscaphe type."

Survey of Mars and Venus Flights*

Name	*Launched*	*Weight*	*Details*
Venera-1	Feb. 12 '61	643·5	Objective: fly-by of Venus at short distance; measurements. Radio contact lost on Feb. 27 when probe was 3·8 million km from earth. Passed Venus on May 19 at a distance of 100 000 km. Went into solar orbit.
Mars-1	Nov. 1 '62	893·5	Objective: measurements and photographs of Mars. Radio contact lost on March 21, '63 when distance to earth was 106 million km. Passed Mars on June 19, '63. Went into solar orbit.
Venera-2	Nov. 12 '65	963	Objective: Measurements and photographs of Venus from a distance of max. 400 000 km. Contact lost when probe passed planet at a distance of 24 000 km (Feb. 27, '66). No photographs received. Went into solar orbit.
Venera-3	Nov. 16 '66	960	Objective: soft landing and information. The spherical capsule was the first to reach Venus, on Mar. 1, '66. Transmission failed on entry into planet's atmosphere. Crash landing.
Venera-4	June 12 '67	1106	Objective: soft landing and measurements. The sphere (383 kg) made a parachute landing on Oct. 18, '67. Information transmitted during descent. Transmission failed at a height of approx. 26 km.

*See also the table relating to Zond 2, p. 222.

Venera-5	Jan. 5 '69	1130	Objective: soft landing and measurements. Sphere (405 kg) made parachute landing on May 16, '69. Contact lost at height of approx. 20 km. Information transmitted during descent.
Venera-6	Jan. 10 '69	1130	Ditto. Landed on May 17, '69.
Venera-7	Aug. 17 '70	1180	Objective: soft landing and measurements. Spherical capsule made parachute landing on Dec. 15, '70. Information transmitted during descent. Contact lost 35 minutes after entry into atmosphere; later restored. Information transmitted from surface for 23 minutes.
Mars-2	May 19 '71	4650	Went into orbit round Mars on Nov. 27, '71. The descent module, previously disconnected, made a crash landing. Parent module continued to transmit signals. Trajectory between 1380 and 25 000 km from Mars. Orbiting time 18 hrs.
Mars-3	May 28 '71	4650	Went into orbit round Mars on Nov. 2, '71. Descent module made soft landing, but transmitted information for only 20 seconds. Parent module continued to transmit information. Trajectory between 1500 and 190 000 km from Mars. Orbiting time 11 days.
Venera-8	March 27 '72	1180	Objective: soft landing plus measurements. Successful landing on July 22, 1972. Information transmitted from surface for 50 mins.

A FIRM FOOTHOLD IN SPACE

Salyut: the first space station – Tsiolkovsky: controllable climate – Space station at base – Specialised space stations – A year in a space dwelling – Building with a box of bricks – Soyuz as transporter – Russian space-shuttle?

"There seems to be no end to this station!"

With this enthusiastic exclamation cosmonaut Georgi Dobrovolski (43) on June 7, 1971 took possession of the very first space station, the Salyut. Commander Dobrovolski and his comrades Vladislav Volkov and Viktor Patsayev had just transferred from their spaceship Soyuz-11 which had brought them safely to a space dwelling orbiting at nearly 200 km above the earth. The Salyut, which provided nearly 100 cubic metres of living quarters, had been launched on April 19 and had been unoccupied for nearly two months.

For nearly 23 days Georgi, Vladislav and Viktor—their radio code-names were Yantar (amber)-1, Yantar-2 and Yantar-3—lived and worked in their cosmic house, thus breaking the record of Andrian Nikolayev and Valeri Sevastyanov, who had travelled in space for 17 days and 7 hours. But that was not all. They also did a great deal of useful work. They made scientific observations. They tested their Salyut. Above all, they proved that well-trained and well-equipped cosmonauts need suffer no ill effects from a prolonged stay in weightless space.* And then, half an hour before the end of the flight, shortly before Soyuz-11 started to dive into the earth's atmosphere—by now an almost routine operation—the cheerful voices of the first dwellers in space suddenly ceased to be heard. The air had escaped from their

*This provisional conclusion is made on the basis of the detailed medical information collected during the flight. On this occasion it was, of course, not possible to judge the cosmonauts' power of adaptation after their return.

capsule and the cosmonauts had died of an embolism, i.e. caused by air bubbles in the blood.

Soyuz-10

The flight of the Soyuz-11 with its tragic ending had been preceded two months earlier by the voyage of the Soyuz-10, manned by the veteran cosmonauts Vladimir Shatalov and Alexei Yeliseyev, and the newcomer Nikolai Rukavishnikov. The significance of this operation was difficult to gauge. Soyuz-10, launched on April 23, 1971, made a flight of only two days. On April 24, at 4·47 Moscow time, Shatalov docked his spaceship into the Salyut. The two craft remained linked for exactly five hours and thirty minutes. Then the crew of Soyuz-10 returned to earth without having transferred—as might have been expected—from one capsule to another.

After the flight the Russians insisted that masses of work had been done and that new rendezvous and docking procedures had been tried out. This was a fact, for the Soyuz was the first spacecraft with a working space equipped with a docking collar, making it possible to change over inside. But if all these systems had functioned correctly, why had no one entered the Salyut?

"No," the cosmonauts Sevastyanov and Popovich told me a few days before the launching of Soyuz-11, "that was not part of the flight programme." When I pointed out to Popovich that the Americans intended to enter their space station Skylab the first time they linked with it, in 1973, he laughed and said: "In the Ukraine we have a good proverb: '*Nye kaziel hop poka nye pereskotsyes*'—don't shout 'hoop-la' before you jump."

Nevertheless it looked as if Shatalov and his colleagues had failed to "jump" only at the last minute. For what was one to think of the flight objective announced in the first Tass communiqué: "to carry out medico-biological research into the effect of factors of space travel on the human organism?" A two-day flight would hardly disclose a great deal in this respect. Moreover, Shatalov had mentioned before the start that he and his fellow crew members had toured the Soviet Union by car for a month, to see if they would get on with each other in space. . . .

"*Yantar, here Zarya....*"

"Yantar, here Zarya. Are you receiving me?"

"Zarya, here Yantar-1. Am receiving you loud and clear."

A conversation between two cosmonauts: Alexei Yelisyeyev in the flight control centre on earth, and Georgi Dobrovolski in his Soyuz-11 which has been speeding round the earth for two days already. The spaceship's automatic pilot has brought the crew to within 6 kilometres of the Salyut.

"Zarya, here Yantar. I have sighted the objective as a small dot!"

The automatic rendezvous system has been switched on. Minutes pass.... Now the Salyut, sparkling in the sunlight, is only a hundred metres away. The moment has come for Georgi Dobrovolski to take over manual control for the more delicate adjustments.

At that moment the spaceship and the space station disappear below mission control's radio horizon. Somewhere on the other side of the earth the two craft dock together and their electrical and hydraulic systems are connected.

Salyut's 796th orbit. The space combination once more appears above the horizon.

The cosmonauts have levelled the pressure between Soyuz and Salyut.

"Zarya, this is Yantar. Pressure in compartment normal."

"Understood. Give the order to open the hatch."

"Zarya, the hatch is open!"

"Yantar, complete the programme. Good luck!"

It is 10.45 Moscow time on June 7, 1971. Dobrovolski, Volkov and Patsayev float into the Salyut. To the men who had been incarcerated in the narrow confines of the Soyuz cabin, their new dwelling seems enormous....

Immediately after entering the Salyut, the cosmonauts start taking the apparatus "out of mothballs". Most of the equipment has been inactive during the two months' unmanned flight. As the Soyuz-Salyut combination, which weighs 25 tons, makes one orbit after another, a lively conversation is maintained between the cheerful crew and mission control. When the spaceship is out of reach of the flight control centre, there is still regular radio contact via the communication ship *Akadyemik Sergei Korolyov* and a Molniya communications satellite.

29. Venera-7 being tested before launching. *Left* The landing capsule. *Smaller plate* Venera landing capsule.

30. Georgi Dobrovols Viktor Patsayev and Vladis Volkov waving goodb before the start of th mission in Soyuz-11.

(*Below*) The space station Salyut in the assembly shed. From l. to r.: docking apparatus; solar cell panels in folded position; astronomical apparatus; far right (in the cut-away section) observation equipment. The large surrounding collar is used only during the assembly, enabling the space station to turn on its axis.

On June 8, Dobrovolski and Volkov, sitting side by side in front of the central control panel, carry out the first course correction. The Salyut has come dangerously close to the upper layers of the earth's atmosphere and unless something is done about it, increasing friction would soon make the station return to earth. A second course correction follows on June 9.

"Zarya," Georgi Dobrovolski reports, "the engine burned for exactly 73 seconds." Patsayev has been unaware of what's been going on. He is "floating" comfortably in his sleeping bag, suspended in the compartment linking the Salyut work space and the Soyuz.

"Yantar, how are you coping with weightlessness?" the earth wants to know.

"Very well. The 'penguin' suit makes a lot of difference," Dobrovolski reports. "It's a first class piece of work. Thank the boys who made it. . . ."

The penguin suit subjects the muscles to a strain almost equal to that on earth and helps to prevent the threatening decalcification of the bones. The cosmonauts wear it nearly all the time. The suit got its name because of the strange waddling movements the cosmonauts made during training on earth.

It takes nearly four days before all the apparatus on board has been put into action and a start can be made with the technical and scientific operations. One of these is testing the wide-angled sextant, which has a range of 190 degrees and may be of great assistance in navigation. The sextant makes it possible to take the altitude of two widely separated stars simultaneously. "It's a great instrument and easy to work," Patsayev, the engineer, reports.

Life on board the space station now falls into a certain routine. The cosmonauts have decided to work a shift system, so that two men are always awake while the third sleeps. They take turns in the extensive technical, meteorological, geological, astronomical and medical experiments. But the programme of eight hours work, eight hours sleep and eight hours off duty remains unaltered. The latter period is used for communication with earth as well as for personal activities, such as reading, or listening to favourite music from the tape-recorder. Then, of course, there are always those daily tasks such as washing.

The Salyut does not have a shower, so the men have to

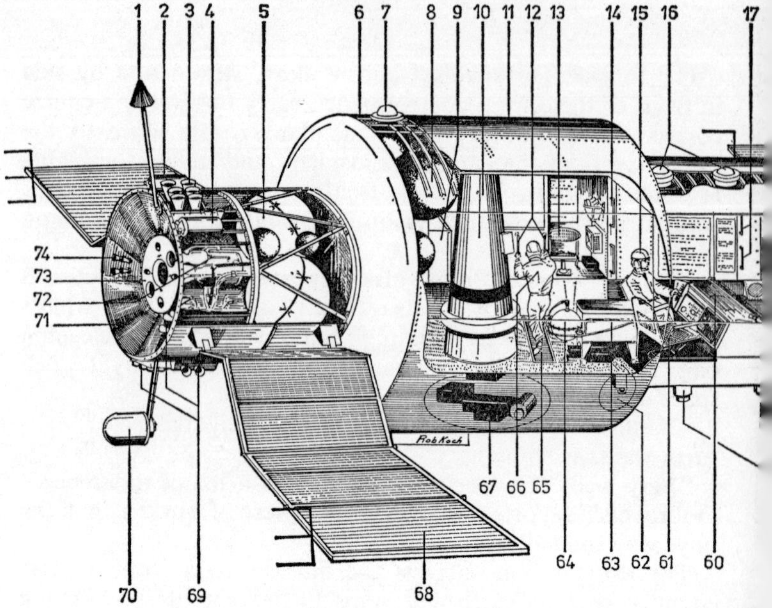

Fig. 29. Salyut orbital station and Soyuz.

1. Homing antenna space station.
2. TV-camera space station.
3. Micro-engines for attitude control.
4. TV-camera space station.
5. Fuel tank.
6. Part of service module.
7. Refilling point.
8. Tank of life support system.
9. Control panel service module.
10. Chimney-shaped section containing equipment, connected with equipment bay 67.
11. Work compartment.
12. Handrails.
13. Chair for medical examination.
14. Handrails.
15. Commander (left seat).
16. Portholes.
17. Handrails.
18. Hatch.
19. Beacon light.
20. Astrophysical system "Orion".
21. Spherical compressed gas containers.
22. Porthole.
23. Mechanism antenna mast.
24. Attitude control jets.
25. Navigation antenna (used during rendezvous).
26. Spherical gas container.
27. Docking tunnel space station.
28. Docking tunnel transport craft.
29. TV-camera Soyuz.
30. Porthole orbital module Soyuz.
31. Navigation antenna (used during rendezvous).
32. Hatch parachute system command module.
33. Parachutes (1 stabilisation, 1 main, 1 back up).
34. Seat (3).
35. Radiators temperature control system.
36. Skirt of propulsion section.
37. Homing antenna.
38. Solar panels (2 on Soyuz).
39. Antenna (communication).
40. Soyuz service module.
41. Sensor orientation system.
42. Telemetry antenna.
43. Instrument panel.

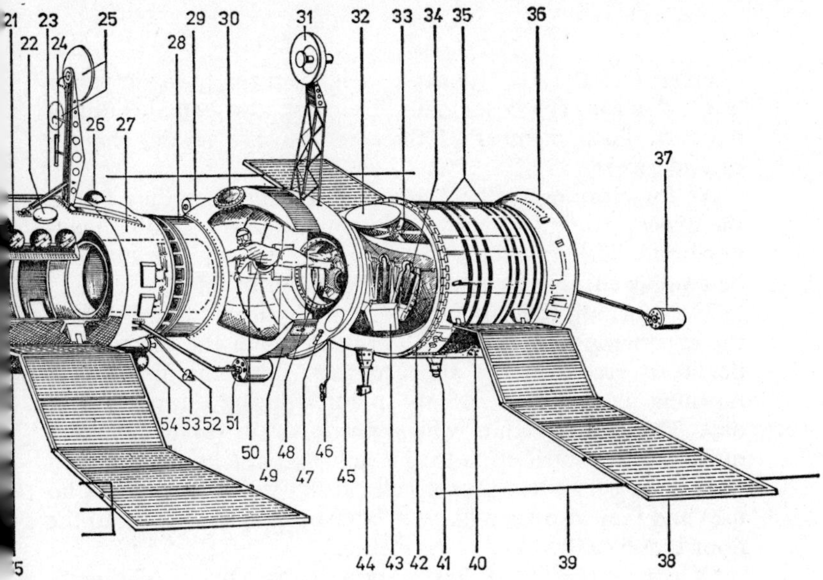

44. Optical orientation device (used during manual rendezvous and docking).
45. Soyuz command module.
46. Beacon light.
47. Stabilisation jets orbital module.
48. Command module hatch.
49. Soyuz orbital module.
50. Cosmonaut transferring from Soyuz to Salyut, holding handrail. (During manned flight of space station, Soyuz command module is sealed and powered down.)
51. Homing antenna space station.
52. Omni-directional antenna.
53. TV-camera.
54. Spherical gas container (see 26).
55. Transfer compartment Salyut.
56. Sleeping bag.
57. Outer protective layer.
58. Equipment section between living quarters and outer shell of work compartment.
59. Sensors and observation equipment.
60. Central instrument and control panels.
61. Seat of flight engineer.
62. Porthole with equipment behind.
63. Table with food containers.
64. Drink water tank with 3 hoses.
65. Treadmill.
66. Cosmonaut doing physical exercises.
67. Instruments in equipment bay.
68. Solar panel space station (4).
69. Stabilisation and attitude control jets.
70. Search antenna space station.
71. Protective skirt of thrust unit.
72. Combustion chambers of 3 main engines.
73. Exhaust nozzles of three main engines and 4 attitude control jets.
74. Attitude control jets.

manage with the wet and dry towels from the hygienic packs, one a day for each cosmonaut. The pack also contains dinner napkins. Each member of the crew also has several changes of underwear.

As for cleaning teeth: here too, weightlessness has forced the experts to find a special solution, namely a kind of fruit-flavoured chewing gum with a cleansing action. The stuff can be swallowed, thus avoiding one pollution problem.

The morning toilet is followed by physical exercises. With the experiences of the Soyuz-9 crew in mind, the cosmonauts devote a great deal of time to this: forty minutes in the morning plus another seventy minutes during the rest of the day. The penguin suits which put a strain on the muscles make a real contribution to this activity. In addition the men work with expanders (dumb-bells, being weightless, are of no use) and they also keep fit on a "conveyor strip" built into the floor in the centre part of the station.

Of course the Salyut has a lavatory. Specimens of excreta will be taken back to earth to be examined.

The fifth day of the flight has been reserved mainly for medico-biological research, and that with reason. Earlier flights have shown that as a rule, cosmonauts adapt to life in space after four or five days. The comprehensive medical equipment on board the station makes it possible to make electro-cardiograms and -encephalograms. Special apparatus measures the density of the bones within a few seconds—this with a view to the calcium content. Salyut's "robot doctor" can take blood samples as well. And of course information on pulse, breathing and blood pressure is continuously relayed to earth. Initial results indicate that physically the cosmonauts show no signs of weakening.

Purely biological experiments relate to animals in conditions of weightlessness. Three hours before launching a sterile container with drosophyla flies was put on board the Soyuz-11. These insects multiply extremely rapidly, so that the effects of weightlessness on subsequent generations can be studied.

Similar tests are carried out on frogs' eggs, which have also been brought along by the Soyuz. By the fifth day they have turned into tadpoles. In that stage of development they are frozen by Dobrovolski. Biologists on earth are interested to see how the balancing organ of these creatures—formed in zero-gravity—have developed. The balancing organs of men and of frogs have much in common. There are also plants on board, such as cabbage, flax and leeks. Dobrovolski reports that the seeds have germinated. Every ten minutes the growth of these weightless plants is recorded on film.

Meanwhile Patsayev is engaged in important astronomical research. Using a special type of telescope, which has been baptised "Anna", he registers the gamma radiation of various cosmic sources. One of the designers of the apparatus, the scientist A. N. Galper, has called the telescope after his eight-year-old daughter Anna, "because, like my little daughter, gamma-astronomy has still a great deal to learn." Patsayev also regularly uses a number of telescopes specially designed for research into the radiation of stars which cannot be observed from earth. This miniature astro-physical observatory is called "Orion" after the well-known constellation.

From time to time the cosmonauts conduct a "guided tour" of their cosmic home for television. On earth it can be clearly seen that Volkov and Dobrovolski are growing their beards. Patsayev does not keep up with fashion: he shaves every day.

A very important object of research is the earth. Photographs are made of the cloud cover, of geological formations, of typhoons and cyclones. Some of these photographs will subsequently be compared with those made by Meteor satellites, thus improving the interpretation of this kind of satellite photography.

On two occasions the cosmonauts take a day off: on June 17 and on June 20. Patsayev's 38th birthday on the 19th is, of course, a good reason to celebrate. Early in the morning he receives warm congratulations from "the earth". "What are you having for your birthday dinner?" mission control wants to know.

Patsayev replies: "Tinned veal, juice from tubes, tongue, Russian cheese, yoghurt, candied peel, prunes and nuts. A copious meal. The only thing we can't bear is that onion cut in three."

Although as a rule conversation is very much to the point, the cosmonauts always have time for a joke. Once Dobrovolski informs Zarya (the earth) that one of the experiments planned has not yet been carried out.

"Why not?" asks earth.

Dobrovolski: "Volkov has been unable to mend the holes in his socks." After the flight of Soyuz-9 Sevastyanov had said that the holes in his socks were the result of his always stubbing his toes when he was floating weightless in the Soyuz work space.

The cosmonauts continue to be in excellent condition. Shortly before the end of the flight, Dobrovolski's pulse is 62 and his blood pressure 115/70, while Volkov's are 64 and 115/68 respectively and Patsayev's 72 and 115/75.

By June 29 the flight programme has been concluded and the three space dwellers prepare for their return to earth. They transfer all the scientific material, the films and the log-books to the Soyuz.

At 9.28 p.m. Moscow time the Soyuz is disconnected from the Salyut, which appears to float away slowly. The spaceship makes three further orbits before starting to descend.

"Here is Yantar-1," says Georgi Dobrovolski. "Everything on board in order. We are feeling perfectly fit. We are ready for landing. "A few seconds later mission control hears a shout of joy from Vladislav Volkov. "I can see the station," he cries. "It's sparkling in the sunshine."

"So long, Yantar, see you soon on earth," says mission control.

"So long. I am starting the landing procedure," Dobrovolski reports. They are the last words received on earth. . . .

At 1.35 that night the retro-rockets are fired. They function perfectly. But nothing further is heard from the crew.

In the flight control centre tension increases. But as yet all hope is not lost. It seems impossible that after this 24-day space marathon, perfect in every respect, a routine operation such as return to earth should be fatal.

"Yantar, here Zarya. Please answer."

"Yantar, here Zarya. Please answer."

But there is no reply from space.

The recovery team has been alerted. Half an hour after the burn of the retro-rockets the capsule lands exactly in the predetermined spot in Kazakhstan. Two recovery helicopters land at the same time. Before the rotors cease to revolve, doctors jump out and race towards the capsule. The hatch is opened. Georgi Dobrovolski, Vladislav Volkov and Viktor Patsayev are sitting lifeless in their seats. Beside them, perfectly arranged, is the scientific material they have brought from their cosmic residence. . . .

The earliest reports reaching the West gave the impression

that communication with the crew of Soyuz-11 had been lost during the period of aerodynamic braking in the atmosphere. It seemed an obvious conclusion that—after their prolonged stay in space—they had been unable to withstand the forces of deceleration. But the original communiqué issued by Tass indicated that the three men had died in space and not in the atmosphere. The communiqué said: *At 1 hr. 25 mins. on June 30, after the spaceship Soyuz had been properly aligned, the braking rocket was fired in order to implement the return to earth. When the braking rocket had completed its function, communication with the crew was lost.*

Then followed a new paragraph: *In accordance with the flight programme, the parachute system went into action on conclusion of the deceleration process in the atmosphere.*

Soon it was unofficially reported that the cosmonauts had died as a result of an airleak in their cabin. The drop in pressure released bubbles in the blood (chiefly nitrogen), which causes blockage in the circulation (an embolism). If—as is the case in American spaceflights—the cabin had contained oxygen instead of normal air, the result would have been the same; in that case too the men would have died of acute lack of oxygen. The only thing that might have saved them would have been a combination of space suits (which they were not wearing; American astronauts do not wear them either during return to earth) and pure oxygen, and then only if they had begun to breathe in oxygen at a much earlier stage, in order to eliminate the nitrogen from their blood.*

In any case the three astronauts can hardly have realised what was happening.

A day after the disaster the bodies of the three cosmonauts were lying in state in Moscow. The fact that the bodies had been released so soon indicated that the cause of death must have been easy to establish. An official statement was not immediately issued. Thousands of Russians, as well as foreigners (including the American astronaut Tom Stafford), paid their last respects to the cosmonauts. The entire world mourned with the relatives: with the cosmonauts' parents, with Ludmilla Dobrovolski and her daughters Maria (12) and Natasha (4), with Ludmilla Volkov and her 13-year-old son Vladimir, with Vera Patsayev and her children Dimitri (14) and Svetlana

*See also: the chapter "Giant Steps in the Infinite", p. 134.

(9). The ashes of the cosmonauts were immured in a wall of the Kremlin.

At last, on July 12, 1971, the special government commission which had been appointed to establish the exact cause of the disaster, issued the following official statement:

FROM THE GOVERNMENT COMMISSION FOR INVESTIGATION INTO THE CAUSE OF THE DEATH OF PILOT-COSMONAUTS OF THE USSR, G. T. DOBROVOLSKI, V. N. VOLKOV, V. I. PATSAYEV.

Examination of the flight recording of the spaceship "Soyuz-11" has established that up to the time of descent, the flight of the capsule took a normal course. The cosmonauts G. T. Dobrovolski, V. N. Volkov and V. I. Patsayev acted in accordance with the flight programme.

During the descent of the spaceship, 30 minutes before the landing, pressure in the return capsule dropped rapidly, which led to the unexpected death of the cosmonauts. This has been confirmed by medical and pathological-anatomical examinations.

The drop in pressure was the result of failure of the hermetic sealing of the spaceship.

An inspection of the capsule, which made a soft landing, has shown that there was no damage to its structure. Technical analysis indicates that there are several possible explanations of the de-sealing. Investigation into the exact cause continues."

Already during the funeral the unofficial explanation had been suggested that the hatch had not been properly closed. This explanation was fairly readily accepted in the West, since it seemed such an obvious one. In the minds of the general public the hatch of the Soyuz-capsule was compared to a car door which had not been entirely locked. The cosmonauts might have "forgotten" to close the hatch properly. But it was not as simple as that.

In September, two months after the disaster, I had a conversation with the medical cosmonaut Boris Yegorov, who had flown in Voskhod-1. For the first time it became clear that the disaster had occurred within a very few seconds, when the work compartment and the command module of the Soyuz had separated. At that moment the air had escaped from the cabin. Yegorov rejected the theory that the hatch had not been properly closed. "The hatch was perfectly closed and bolted," he said, "but there are several connections, including pneumatic and electrical ones between the work compartment and thc command module. A failure must have occurred in these connections." Yegorov declared frankly that it was very difficult to determine exactly where the failure had occurred, the more so as the work compartment was not available for inspection.

From Yegorov's words it might be inferred that one of the valves equalising the pressure between work space and command module had continued to function after the two sections had separated, which would have meant that the pneumatic connections would have opened onto the vacuum of space.

Although my interview with Yegorov was published in several journals, the story about the imperfectly closed hatch continued to circulate, even after Mstislav Keldysh, the President of the Academy of Sciences, had made a statement which concurred with the theory of valve failure. Perhaps this was because the statement formed an unemphasised part of a long speech, which Keldysh made in the Kremlin Congress Centre on the occasion of the Day of Cosmonautics, on April 12, 1972. He said: "The entire Soviet people and the entire world were deeply saddened by the sudden death of the cosmonauts, when, after they had successfully carried out their flight programme, one of the elements of the pressurising system unfortunately failed during the descent of the spaceship Soyuz-11 to earth."

Salyut-1 was not manned again. After a number of technical experiments by remote control, the station was decelerated on October 11, 1971, on a signal from earth, after which it burned up in the denser layers of the atmosphere. Salyut had functioned perfectly for nearly six month on end. It was obvious that in the future the personnel of such stations could be repeatedly relieved.

The flight of Salyut and Soyuz-11 had not been in vain. As had been planned, more and more cosmonauts would be able to live for increasingly prolonged periods in space in the seventies and eighties. . . .

Soyuz-12 Testflight

"Why dream? It is a job, my job. It has become my flesh and blood. Even if you have not flown in space, you inevitably become a professional. As for this flight, to be frank, I didn't have time to feel that the start was drawing near, being preoccupied with things at hand . . . especially as Oleg and I realised that our flight was not to be a super achievement. Our tasks were purely technological, in other words, our everyday duties, which we simply happened to be doing in a different environment."

This is how cosmonaut, test pilot and doctor Vassili Lazaryev (45) described the short but very important flight of Soyuz-12; a flight in which he and Cosmonaut Oleg Makarov (40) flight-tested an improved Soyuz for two days. This was almost 27 months after the Soyuz-11 crew had left the earth on its ill-fated cosmic flight.

During the flight of Soyuz-12 Makarov, an engineer employed in the design-bureau, had an excellent opportunity to evaluate the various new systems of Soyuz, especially the environmental control system. Makarov has been involved in the design and development of manned space craft ever since work on Vostok started.

Soyuz-12 was launched on 27th September, 1973. For the first time Tass announced shortly after lift-off how long the flight was going to last. The flight was characterised by extensive safety measures. Both cosmonauts wore space-suits during the critical phases of lift-off and re-entry. And before separation of the orbital and command modules at the end of the braking manoeuvre, the cosmonauts tested the air-tightness of their capsule.

During the flight they also did extensive manoeuvring including changes of their orbital path. They also took many photographs of the earth in various bands of the electromagnetic spectrum, destined to be used in studies of the earth's resources. Some biological experiments were also performed.

"The results of the flight are to the complete satisfaction of all involved in this experiment", said Vladimir Shatalov (chief of cosmonaut training). The flight was welcomed by NASA-officials as a demonstration of the reliability of the improved Soyuz, which is to be used for the joint Apollo-Soyuz experiment in 1975.

The flight of Soyuz-12 also opened up new possibilities of prolonged flights on board the Salyut space stations.

Why take such risks?

The idea that human beings should spend prolonged periods or perhaps even their entire life—in space, is at first sight somewhat absurd. Man's evolution has taken place on earth; he is entirely adapted to its ecological conditions. He cannot live without a pressure of approximately one atmosphere, nor without oxygen, water and food, or a suitable temperature. Therefore, if a man wants to live, or even spend only a short time, in space, he can survive only by surrounding himself by all manner of complicated apparatus. Why, sceptics ask, should we take all this trouble, when we can exist on earth without all such technical equipment?

Konstantin Tsiolkovsky gave the answer to this question in his book "The Purpose of Space Travel", which was published at the beginning of this century. He wrote: "Of course the absence of air or atmosphere (in space) is the most disagreeable factor. This is, however, only partly true, for the atmosphere is at the same time a source of misery for mankind. Man is unable to control the atmosphere, its temperature or any of its other aspects. Take temperature, for instance. On the equator the heat during the day makes life almost impossible; at night it is more bearable, but then the air is damp and unhealthy. In northern countries the summers are unbearably hot and the winters equally unbearably cold. What sacrifices and exertions man has to make in his struggle against temperature, against wind, snow, downpours, droughts, germs, etc. The atmosphere, moreover, robs mankind of a large part of solar energy: one part is reflected by the clouds, another is absorbed by cloudless air. The atmosphere is a thief!"

This loaded statement may be somewhat exaggerated, but objectively considered it is not far out. Tsiolkovsky says with reason: In a space station we can arrange conditions as we want them. Viewed in this way, the earth is a spaceship in which temperature, air circulation and other conditions are entirely incalculable.

Cities in the ether

The concept of "cities in the ether", in our phraseology space stations, forms the nucleus of Tsiolkovsky's vision of space travel. Tsiolkovsky considered the construction of a permanent base in space to be the first important step. Only after this had been achieved could, for instance, a landing on the moon be undertaken. Later still the nearer planets could be explored. This concept characterises the Russian space programme. Hence the Soviet Union concentrates its activities in space on the establishment, first of temporary, and later of permanent bases in space; space stations capable of housing dozens and even hundreds of people. Such space stations can serve as a springboard for travel further into space as well as being of enormous economic and scientific value.

There are numerous possible applications of a space station. Naturally the most obvious is its use as a launching base, but in addition it is not difficult to mention half a dozen further areas in which it can be used, for instance meteorology, oceanography, geology, astronomy, economics, biology.

The idea of using a space station as a starting point or intermediary port of call for interplanetary spaceships is a logical one. Spaceships can start from them without difficulty on their way to the moon and the planets. The station itself already orbits the earth at a speed of eight kilometres a second. A spaceship starting from such a base need develop only an additional three kilometres per second to reach escape velocity. Moreover, on its return the spaceship can dock at the station and need not be subjected to the rigorous conditions of re-entry into the earth's atmosphere. The same spaceship can therefore be used again and again. For communications between the earth and the space station cheap spaceships, built in series, such as the Soyuz, can be used, or, later on, a space shuttle which can take off from and land on any normal airfield. This too, can be used again and again. In this way a permanent space transport system can be created which does not involve the repeated and enormous cost of "obsolescent" rockets and spaceships.

The Americans admit that the Russian view is the right one. They cut down their Apollo programme, which has cost them thousands of millions of dollars, from the twenty flights originally planned to seventeen, in order to spend their budget on the development of a space shuttle and a space station. The first, modest, American space laboratory (Skylab) was launched in 1973. The fact that the Soviets indeed regard space stations as starting points for more distant flights, was confirmed by space walker Leonov, when I asked him: "Do you think that the space station will also be used within the framework of interplanetary flights?"

Leonov laughed, and said: "It would be strange if I answered in the negative. Of course I think that will happen."

I asked: "Will Soviet cosmonauts set off to the moon from such a station or do you think that lunar exploration is better done by robots?"

Leonov: "Of course. To the moon, to the planets. Where you will!"

A space station also makes an ideal meteorological institute. Seated, as it were, "in the stalls", meteorologists will be able to observe the distribution of cloud formations and of areas of high and low pressure with their own eyes, thus getting an overall picture unobtainable on earth or just above it. The cosmic centre will continuously relay photographs and details of the distribution of cloud formations, barometric pressure, whirlwinds, the boundaries of snow and ice cover. "And just think of the opportunities which the station offers to geologists and cartographers," Cosmonaut Vladimir Shatalov said enthusiastically, "not to mention agriculturalists. The growth of crops can be watched from space. In this way it will also be possible to give early warning of certain plant diseases. We shall, moreover, be able to trace ocean currents and take temperature readings. This information, in turn, can indicate the presence of large schools of fish. The value to agriculture, shipping and many other economic activities of the meteorological information supplied by a space station can hardly be overestimated. Long-term weather forecasts with a reasonable degree of accuracy will become possible for the first time."

Astronomers get quite excited when they talk about the opportunities a space station will offer them. Unlike his colleagues on earth, an astronomer will, from a space station, be able to get an unimpeded view of the heavenly bodies. For centuries astronomers have regarded the earth's atmosphere as an enormous obstacle. In the first place all manner of atmospheric pollution clouds the view in the telescope. The atmospheric turbulence caused by whirlwinds and air currents is also a great handicap. This turbulence, for instance, causes the moon, when seen through a greatly magnifying telescope, to appear as a pancake in the process of being fried, making it impossible to distinguish details.

Gravity, moreover, limits the size of our telescopes. Telescopic object-glasses cannot, in practice, exceed one metre in diameter, since they are supported only round the edge and cause distortion as a result of their own weight. Reflecting telescopes are not quite so subject to this limitation, but in order to prevent distortion, the mirror has to be supported by a very heavy structure, so that this kind of telescope weighs hundreds of tons. In space all these problems are avoided. There is no air, so no dust and no turbulence. Nor is there gravity, so in theory at least, instruments can be as large as may be desirable.

The earth's atmosphere forms a considerable obstacle to astronomers in another way as well. It allows only a very small part of the electro-magnetic radiation emanating from space (including visible light) to penetrate to earth. This is just as well, for some forms of radiation would put an end to all forms of life on earth. To astronomers, however, it is disastrous. The range of wavelengths which do penetrate the atmosphere form, as it were, only a few small "windows". Astronomers on board a spaceship will not have to look at the universe through these small "windows"; they can survey the cosmic landscape as a whole, without interference.*

The space station is exceptionally suitable as a laboratory for experiments requiring a complete vacuum, zero-gravity or the presence of certain forms of radiation. Some industrial processes, such as the manufacture of perfect spheres for ball-bearings, could be carried out better than anywhere else. On earth it is impossible to produce a perfectly spherical ball because of the pull of gravity. The Russians attach great importance to the use of space for industrial and other economic purposes.

The absence of gravity and the presence of various forms of cosmic radiation—at least when protection against such radiation has purposely been omitted—is a veritable godsend to biologists, who are anxious to know what effect such conditions have on living organisms (unicellular organisms, plants, animals, man). Experiments with plants and animals have been carried out on board rockets and spaceships, but these have always been of relatively short duration. The biological research carried out on board Salyut introduced an entirely new phase.

*Unmanned astronomical satellites, such as the American OAO, are already a great step forward in the study of the universe.

Specialised space stations

Both the construction and the trajectories of future space stations will depend on the applications envisaged for them. Space stations can orbit close to earth—at a few hundred kilometres, as did Salyut—or at a great distance. They can follow a polar course, an equatorial course, or a trajectory somewhere in between, as did the Salyut and the Soyuz, whose orbits always made an angle of about 51 degrees with the equator.

Orbits close to earth, such as that of the Salyut, are impracticable for space stations which are intended to be "permanent fixtures". At such heights air resistance is still sufficient continuously to decelerate a moving object. A permanent space station could continue to orbit at such a height only if its altitude was corrected at regular intervals by means of an inbuilt rocket engine. On the other hand, space stations at high altitudes would be of little value if their purpose was to study changes on earth—climatological or otherwise—particularly those on a relatively small scale.

There are two types of special trajectories: the polar and the equatorial. Polar orbits are ideal for space stations serving as observation posts, whether for military purposes or not. As the station revolves, the earth turns below its course once every 24 hours, so that each point of the earth's surface comes into view twice every 24 hours.

An equatorial orbit also has specific advantages, particularly where three space stations orbit at a height of 36 000 km at distances of 120 degrees from each other. Such a trio would be ideal for recording global meteorological patterns and also as a relay centre for radio and television signals. They could serve to connect practically all places on earth, with the exception of very small areas round the poles. To the Soviets, however, putting satellites into equatorial orbit presents an impossible task as long as they have no launching base at their disposal near the equator, for it requires enormous forces to transfer a satellite from a course with a normal inclination (50 to 60 degrees) to an equatorial one.

Not only technological but human factors play a major part in the design of a space station. The desirability of producing artificial gravity on board a permanent space base has been mentioned before. This is a purely biological requirement. However, there is also an important psychological problem. How will the crew react to each other if they have to live for months at a time, isolated from the earth and its inhabitants, in a fairly restricted area? Might they become irritable or possibly even aggressive—which could affect their efficiency?

The Russian reply to this question is that no serious problems are to be expected, provided people are selected whose characteristics do not clash.

The Soviets speak from experience, for three Russians have lived on board a "space station" for a whole year already—only, the station was still on terra firma. On November 5, 1967, Dr. German Manovtsev (31), the biologist Andrei Bozhko (29) and the technician Boris Ulibishev (24) entered their quarters, which had a floorspace of only twelve square metres. Exactly twelve months later they emerged. In all that time their only contact with the outside world had been via a videotelephone. Water and oxygen were continuously purified for renewed use; they ate dehydrated food (salmon, chocolate, cheese tablets, prunes—all made digestible by soaking in water) and vegetables which they grew themselves in a greenhouse: cabbages, lettuce, fennel, watercress, etc. These plants also played a rôle in the closed cycle of the ecological system. The experiment was therefore an important way of testing an autonomous system of "environmental viability", as will be applied in permanent space stations.

The walls of the earth-bound space station had been decorated in soft pastel shades, selected by psychologists. Every morning, at seven o'clock sharp, the men got up; they went to bed at midnight. The crew had enough work to occupy them, but of course they had time to watch TV, to read and to play chess.

On February 25, 1968, Manovtsev heard that he had become the father of a fine daughter, but he had to wait nine months before he saw her.

Originally the men had expected that they would not be able to stand living in their confined quarters for more than five months. The fact that they stuck it out much longer did not, however, mean that they had encountered no difficulties. There were occasional problems, particularly in the beginning, but the three men learned to be tolerant, even though this was

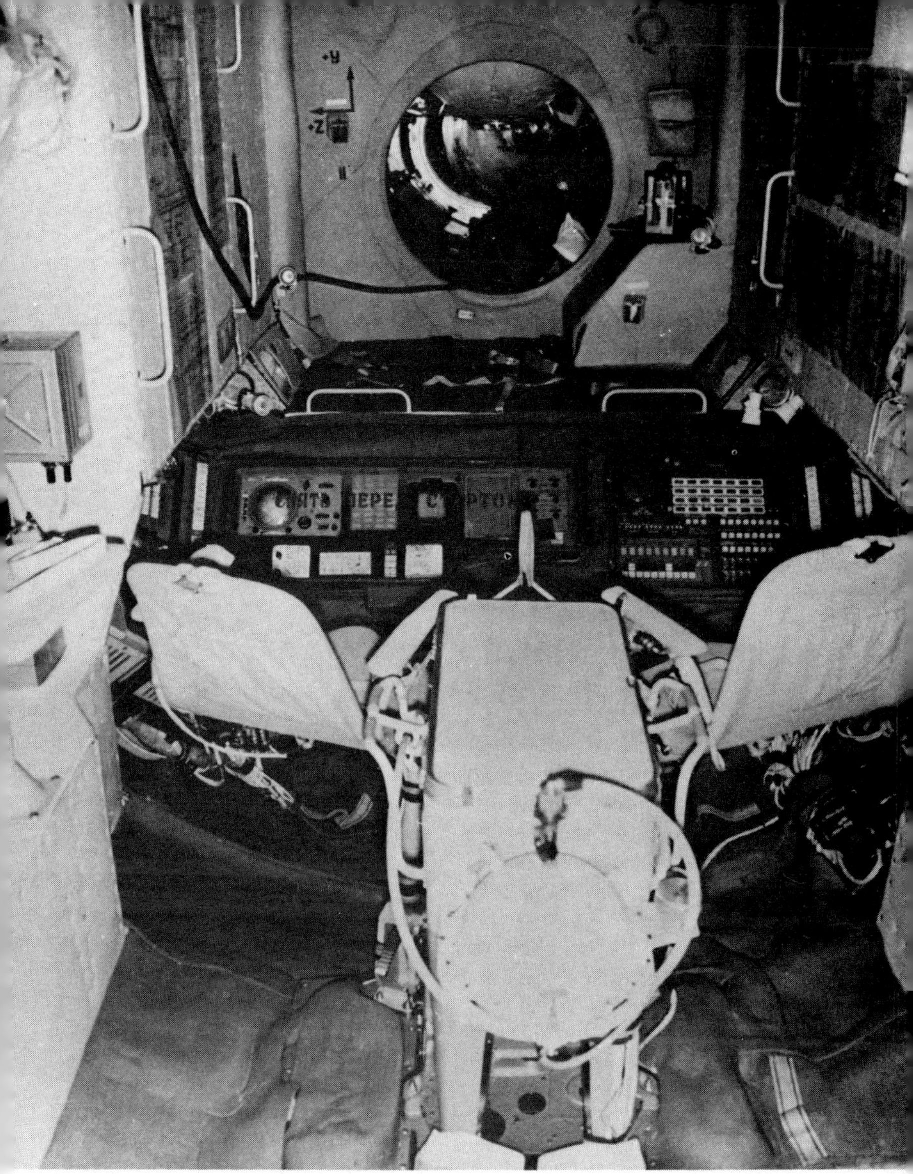

31. Part of the interior of the Salyut. Walls and floors are decorated chiefly in pale blue colours. In the background the hatch to the transfer compartment, which contains, among other things, astronomical apparatus and sleeping bags for the cosmonauts. In the centre the service and guidance control panel and two chairs. Far left, in a niche, the chair for medical examinations. The handrails attached to the walls allow the cosmonauts to hold on when weightless. The inscription on the protective cover reads "remove before lift-off".

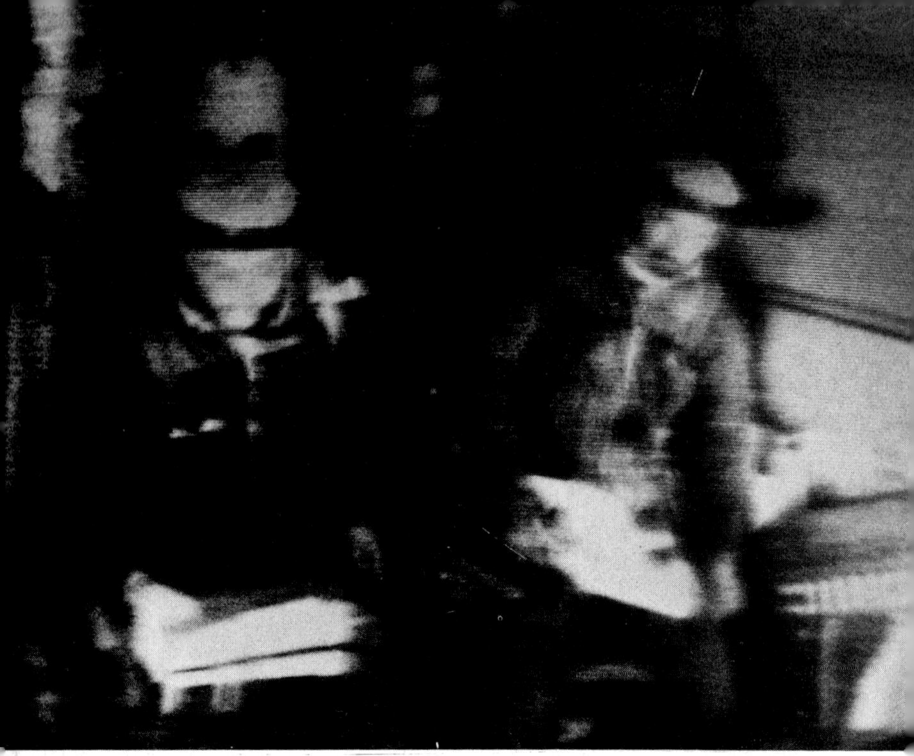

32. (*Above*) Cosmonauts Dobrovolski and Volkov, seated in front of their central service and guidance control panel in the Salyut. (Television picture). Behind them the engineer Patsayev.
(*Below*) The rendezvous of the century: in 1975 a 3-man American Apollo and a two-man Russian Soyuz are to link up in space. A standardised transfer tunnel and docking apparatus have been developed for this purpose. The standardised docking gear will in future make mutual rescue operations in space possible.

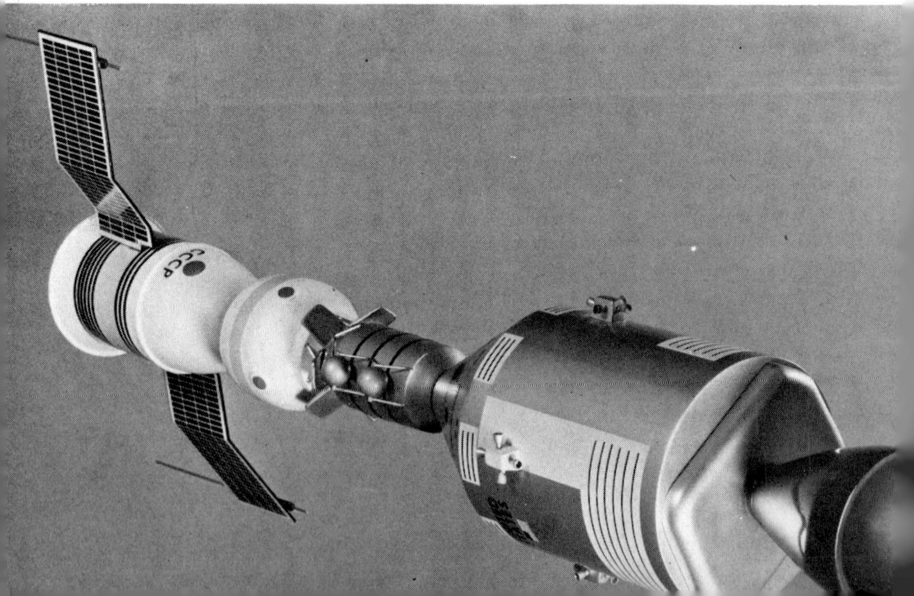

not always easy. Bozhko, the biologist, said: "The isolation was really hard to bear. I particularly missed my friends and acquaintances of the University of Moscow, where I had just finished my studies." Manovtsev said afterwards that, particularly in the first few months, he had regularly felt a longing to take his little daughter for a walk in the city on a rainy night. Ulibishev missed his family most. He, too, occasionally felt like taking a long walk in the country.

The three men said that the final minutes of the experiment had been particularly difficult. All three had been excited and irritable. When on November 5, 1968, the door of their quarters opened at last, they were not immediately allowed to go home—they had to get acclimatised in a clinic first.

The scientists as well as the guinea-pigs themselves were extremely satisfied with both the psycho-biological and the technical results of the experiment. Concerning the psychological aspects Manovtsev summed up his experiences as follows: "We learned fairly soon to avoid acute conflict. Perhaps we have become more tolerant towards our fellow men."

Building on the module principle

Tsiolkovsky, in his day, already described "greenhouses", attached to space stations, which were to supply the necessary fresh vegetables and fruit. The Russians consider that such a cosmic garden may be a possible solution of the provisioning problem on board a large space station, which might have hundreds of residents. In Moscow such a greenhouse is already being tested in a large cylinder, three metres in diameter, which could easily be incorporated into the design of a carrier rocket.

The best form for the various sections of a space station is that of a cylinder. These sections can easily be placed into earth orbit by medium-sized rockets. It is this module system which makes the construction of very large space stations attractive as well as practicable. By putting a large number of comparatively lightweight elements close together in an orbit round the earth, to be assembled there, the need to develop mammoth rockets, requiring the investments of enormous amounts of money, is avoided.

"*Will the cosmonauts themselves play an active part in the assembly of large space stations?*" I asked Cosmonaut Leonov in Moscow.

Leonov: "We are not exactly conventional in this respect. Let robots do the things for which they are suited, and let men do what they are suited for. Of course there are operations which would be impossible or awkward for a robot to carry out. In that case let a man do it. Naturally the assembly of complicated space stations will have to be carried out by human beings. That, after all, is why we travel in space."

I asked: "*What do you think of the possibility of working outside in space? The first space-walks presented some problems in this respect.*"

Leonov: "These problems were primarily due to the imperfections of the spacesuit. There is also the lack of a firm foothold, the floating in space. The latest spacesuits are more flexible and supple and have better ventilation. Spacesuits have, after all, been a subject of research only for the last ten years. I am convinced that all the work we are doing on earth with all kinds of special tools can later be done in space as well."

Nikolayev, the cosmonaut who piloted Soyuz-9 through space for eighteen days on end, envisages a new profession in the area of space travel: that of cosmonaut-mechanic. The elements can be brought together by remote control from earth, but cosmonauts will be required for the finishing touch: tightening screws as well as welding and soldering. For this kind of work special tools are being developed on earth, says Nikolayev. The cosmonaut-mechanics should have a private rocket, so that they can move rapidly on and around the "building site".

Future space stations will be assembled from special elements, all according to the purpose for which they are intended. Apart from residential "boxes", they will have to contain a laboratory and servicing sections. The engineers Anatoly Andanov and Gennady Maximov say that the elements will as a rule be constructed in such a way as to combine minimum weight with maximum volume. They will, therefore, be mostly round or cylindrical. The space station itself can, in principle, have any desired shape. Small space stations for the execution of limited programmes might consist of:

—a command module containing all control and steering apparatus; in this cabin the cosmonauts could return to earth;

—a laboratory for carrying out experiments previously planned;
—an "orbital compartment", containing recreation rooms as well as the necessary sources of energy, supplies and engines for course corrections.

The Salyut is, in fact, of this type.

Practically all forms of space stations have their advantages and their disadvantages, say Andanov and Maximov. A simple cylinder, possibly built up of smaller cylinders, presents no problem as far as its transport is concerned, since it can easily be incorporated in the layout of a carrier rocket. On the other hand, no artificial gravity can be produced in a cylinder, so the station could not be inhabited by the same cosmonauts for long stretches of time. Moreover such a cylinder must constantly be stabilised with the aid of small engines, as otherwise it will tumble uncontrollably.

A large ring-shaped space station can certainly not be launched by one or two rockets only; but such a base, consisting of a large number of cylindrical sections, could be set in circular motion, which would make it very stable. Once such a space giant is rotating, no engines will be needed, since there is no friction. The axis of such a space wheel would constantly maintain its position, like the centre of gravity of a gyrating top. If in addition all floors in the station are made parallel with the axis, artificial gravity is produced which will make the cosmonauts' life much more bearable.

A combination of these two types of space station is one built up of two or more cylinders, far apart, but linked by a "spoke". A central axis can be mounted in the centre of the spoke, where supply spacecraft could dock. The living-, working- and servicing-compartments of the station would be connected by a lift in the spoke. Of course this cosmic giant would also rotate on an axis to produce artificial gravity. The advantage of such a station which, judging by Russian documents and by what Cosmonaut Shatalov told me, might well become a reality, is its relatively simple construction. It may, moreover, be regarded as an extension of a simple cylindrical space station, such as the Salyut.

In addition space stations of the future will undoubtedly have "outbuildings" where special experiments can be carried out, for instance experiments requiring complete absence of vibrations. Such vibrations are bound to occur in the gigantic space platform itself, where so many activities take place.

There may also be experiments involving radio-active materials, which must be kept away as far as possible from the inhabited sections of the station. In the long run, space close to earth will be traversed by space stations of all types, large and small, with and without artificial gravity, with and without "annexes".

A Russian space shuttle?

In the initial phase, transport between space stations and the earth will be provided by versions of the Soyuz adapted for the purpose. "The Soyuz is the spaceship of the future," according to Professor Anatoli Blagonravov.

When I was in Moscow, General Kamanin confirmed that, for the time being, the Soyuz would indeed be used for transporting men and materials to and fro. "It is a very good, reliable spaceship," he said. "It is manufactured in series and is therefore very economical. It can be used for all kinds of jobs."

In the United States spaceships are being developed which are a happy mean between an Apollo and a Concorde—streamlined machines with short wings, which will be able to take off from and land on normal airfields. Unlike the Apollo and the Soyuz, these machines can be used again and again. NASA will use them for maintaining communications with permanent space stations, which will become a reality by the mid-eighties. Although Russian technical journals often contain descriptions of this American space shuttle, there is never any mention of a Russian counterpart. Have the Russians any plans in this respect? In view of their obvious tendency towards economy in the space industry, one might expect this to be the case. I therefore asked Professor Sedov: "Do you think that the USSR will also use space shuttles in the future?" And I showed him a photograph of such an American machine.

"Yes," he said evasively, "that sort of thing might be expected in the future."

I asked: "Are Soviet scientists doing anything in this field?"

Sedov: "Yes, they are exploring all theoretical possibilities."

Sedov therefore admitted that the Russians are studying the space shuttle, which might, in any case, be logically expected. Considering Russia's great achievements in the field of supersonic flight (Tupulov-144) and their obvious endeavours to develop an efficient space transport system, it is more than likely that the Soyuz will bc succeeded by a Russian space shuttle. Once that is in operation the Russians will have the most important element of a system which will turn large-scale space travel into a viable, daily proposition.

Summary

Salyut. *Launched April 19, 1971, 04·34 hrs. Moscow time. Initial weight approx. 18·9 tons. Angle between trajectory and equator: 51·6 degrees. Orbiting time: 88·5 mins. Perigee 200 km, apogee 222 km. Course correction on April 27: perigee 251 km, apogee 277 km. Course correction June 8: perigee 239 km, apogee 265 km. Course correction June 9: perigee 259 km, apogee 282 km. Course correction June 30: perigee 235 km, apogee 295 km. Burnt up in the atmosphere after braking on October 11, 1971.*

Soyuz-10. *Launched April 23, 1971, 02·54 hrs. Moscow time. Initial weight approx. 6·5 tons. Angle between trajectory and equator: 51·6 degrees. Orbiting time 89 mins. Perigee: 208 km, apogee 246 km. Docked with Salyut April 24, 04·47 hrs. Moscow time. Separated after 5 hrs. 30 mins. Duration of flight 47 hrs. 46 mins. Landed April 25, 1971, 02·40 hrs. Moscow time.*

Soyuz-11. *Launched June 6, 1971, 07·55 hrs. Moscow time. Initial weight approx. 6·5 tons. Angle between trajectory and equator 51·6 degrees. Orbiting time 88·3 mins. Perigee 185 km, apogee 217 km. Docked with Salyut June 7, 1971, 10·45 hrs. Moscow time. Separated June 29, 1971, 21·28 hrs. Moscow time. Duration of flight 552 hrs. 17 mins. Landed June 30, 1971, 02·12 hrs. Moscow time.*

Soyuz-12. *Launched September 27, 1973, 15.18 hrs. Moscow time. Initial weight approx. 6.5 tons. Angle between trajectory and equator 51.6 degrees. Orbiting time 88.6 mins. Perigee 194 km, apogee 249 km. Duration of flight 47 hrs. 16 mins. Landed September 29, 1973 14.34 hrs. Moscow time.*

THE FUTURE: CO-OPERATION

Towards a Russo-American space station? – A modest beginning – Co-operation with socialist countries and with France – Inter-cosmos – Standardised docking gear – Rescue operations in space – Space travel: International

"It is my personal belief that a Russo-American space station may well one day become a reality."

"Each flight demands immense financial sacrifices and the problems we face are problems of all mankind. This certainly applies to the exploration of distant planets, which is not a task for one single country."

These, respectively, were the replies of the astronaut John Young (Apollo-16) and the cosmonaut Vladimir Shatalov (commander of Soyuz-10), when I asked them both the same question: "What do you think about possible co-operation between the United States and the Soviet Union in the field of space travel?"

Of course the suggestion that the US and the Soviet Union should work closely together in space is an obvious one. All things considered, it is absurd that this small planet earth should be "represented" in space by two great powers (at least as far as manned space travel is concerned). Space travel transcends the dimensions of our planet. There has, in fact, been talk of joint Russo-American space activities ever since the first Sputnik was launched, but only lately have there been signs which clearly indicate a favourable development of this idea. Up till now co-operation between the USA and the USSR has been restricted to very modest contacts.

The problem of co-operation presents four clearly defined aspects: human, political, technical and organisational. The human side is undoubtedly the simplest. American and Russian space travellers, scientists and technicians get on well together. Several American astronauts have visited "Star Village" near Moscow, where they were allowed to have detailed conversations with their Russian colleagues and to try out Russian training facilities. The same applies to American technicians.

In return cosmonauts visited the manned spaceflight centre in Houston. Soon after their record spaceflight, Nikolayev and Sevastyanov spent ten days in America, where, "for fun", they made a number of simulated moon landings in the LM-trainer. Three times Nikolayev tried to land, assisted by the astronaut David Scott, but according to the training centre's technicians, he "crashed" each time. Sevastyanov made one unsuccessful and one hard moon landing, and quickly pressed the abort-button to simulate an emergency take-off! In an Apollo flight-simulator the two cosmonauts attempted docking with a Soyuz projected on a screen. According to a NASA technician, Sevastyanov was "wide off the mark", while Nikolayev ran out of fuel.

But their American colleagues said that they'd done very well. "Those things (the simulators) are very difficult to use, especially when there are language problems." Language problems can of course be solved.*

If there is to be intensive co-operation, American and Russian space travellers will have to become familiar with each other's technical facilities, but this need not present insurmountable problems either.

The most difficult aspect is the political one. In the past the Russians always said that extensive technical co-operation would be possible only after a disarmament agreement had been reached. Rockets, after all, can be used for launching atom bombs as well as spaceships. It remains to be seen how far the Russians will in future adhere to their old demands, but it is striking that their willingness to co-operate with the Americans is now very marked—as will be seen below. Nevertheless a base such as that at Baikonur remains closed to American space travel experts. The only Westerners who have been allowed to visit it have been De Gaulle and Pompidou, but these visits could hardly be regarded as dangerous from a technical and military point of view. Russian cosmonauts have been invited on several occasions to visit Cape Kennedy, but they have always politely but firmly refused these invitations . . . probably to avoid having to invite the Americans to Baikonur in return. . . .

*During an American press conference on the subject of co-operation in space a journalist raised a laugh when he asked whether astronauts and cosmonauts working together might perhaps speak German. Be that as it may in 1972 a number of American astronauts began learning Russian.

In the field of organisation it would be possible to do everything necessary without getting involved in political and technical problems of any importance. There has never yet been any mutual consultation between the two space powers on how the various tasks in the exploration of space could be divided. Such an arrangement could first be applied to unmanned projects. Both East and West launch scientific earth satellites with the regularity of clockwork, without ever consulting each other. It cannot be remotely estimated how many experiments have been duplicated in this way. Another fruitful field for co-operation may be found in the extremely expensive projects for communications satellites. As early as March 1962, Khrushchev suggested that something should be done in this area, but it has remained a suggestion only, and at the moment both the Americans and the Russians have their own communications satellites.

In spite of this lack of mutual consultation, American and Russian projects fortunately often supplement each other in practice. The Russians are not keen on doing things which the Americans have already done or are about to do. The United States, for instance, have done a lot of work on the exploration of Mars, while the Russians essentially concentrated—and with success—on Venus. The exploration of the moon is a case in point: while the Americans sent their astronauts there, the Russians obtained interesting information by means of their unmanned projects, such as Luna-9, Luna-13, Luna-16, Luna-17 and subsequent Lunas.

Modest contacts

On March 20, 1962, Khrushchev wrote in a message to Kennedy: "Once agreement has been reached on disarmament, much wider perspectives will be opened for co-operation and for joint scientific achievements, including the construction of spaceships for the purpose of reaching the other planets: the moon, Venus and Mars." In the same message Khrushchev pleaded for a start to be made with co-operation on a more modest level: the exchange of scientific information, the drafting of an international agreement on rendering aid to space crews in difficulties, and joint experiments in the field of communications.

Fairly soon afterwards the Russians and Americans made contact on a juridical level. Towards the end of 1961, after consultations between the two nations, in which other countries subsequently took part, a UN resolution had been passed on "International co-operation in the exploitation of space for peaceful purposes." For several years this was all that was achieved, but towards the end of 1966 the US and the Soviet Union reached agreement on a treaty determining the international and non-military status of space. This treaty was adopted by the UN as well. It stipulates, among other things, that no one can lay claim to any part of space or of a celestial body (for instance, the moon), that any nation may explore celestial bodies for peaceful, scientific purposes, and that space may not be used for distributing arms.

Two years later the "Twenty-nation treaty" was concluded, which embraced international co-operation in the rescue of space travellers in emergencies and the return of spacecraft which have landed on foreign territory. The many offers of help in the recovery of the crippled Apollo-13 received by the Americans (from, among others, Prime Minister Kosygin) could be viewed in the light of this treaty.

Modest scientific contact between the US and the USSR culminated in August 1963 in an agreement between NASA and the Academy of Sciences of the USSR, to co-operate in the field of weather satellites, experiments with the balloon satellite Echo-2 for the purpose of reflecting radio-signals, and to make a joint study of the earth's magnetic field, partly with the aid of satellites. In addition the two countries would be able on their own initiative to exchange purely scientific results in the area of space exploration.

As a result of this agreement, dozens of photographs and other meteorological information supplied by weather satellites have been exchanged since 1963 between Moscow and Washington via a special "hot-line". Meanwhile the exchange of other scientific information of a non-technical and non-military nature has also been put into effect. Moreover Russian and American experts are co-operating on a voluminous standard work on the safe subject of cosmic biology. Nor did the Americans hesitate to put the complete and largely technical report on the problems encountered with Apollo-13 at the disposal of the Academy of Sciences.

Following on his correspondence with Khrushchev, Kennedy

on September 20, 1963, again pleaded in the UN for a joint Russo-American manned spaceflight. By then the preparations for the Apollo project were well under way. A few weeks later the American House of Representatives voted 125 to 110 against appropriating funds for such an undertaking. At that time they preferred a manned moon-landing to be a great national operation, rather than a project in which other nations could play a decisive part, even though the project would cost around twenty thousand million dollars.

The lack of contact in technical matters between eastern and western spaceflight experts led to the use of greatly divergent systems. To name but a few: the Americans use pure oxygen, the Russians normal air for the atmosphere of their spaceships; the Americans land on water, the Russians on land; the Americans have an enormous diversity of designs, the Russians concentrate on standardisation. This difference in technical concept presents an additional difficulty in the realisation of a more intensive co-operation between the two countries, although the problem need not be insurmountable.

Intercosmos

Ever since 1957, the year in which the first Sputnik was launched, the Soviets have co-operated with spaceflight specialists from other socialist countries. At first this co-operation was restricted mainly to observation posts in the various countries where the trajectories of Soviet satellites were recorded; but since April 1967 a concrete programme for space exploration has been developed, in which Bulgaria, Hungary, Poland, Rumania, East Germany, Czechoslovakia and the Soviet Union participate. Meetings of scientists of these countries now take place regularly, usually headed by Professor Boris Petrov, the Chairman of the Commission for International Co-operation in the Field of Cosmic Space Exploration.

Joint geophysical experiments were carried out with Kosmos-261, launched on December 20, 1968. After these introductory tests, the six countries co-operated in the design of new artificial satellites, which were launched under the name of "Interkosmos". The apparatus for these satellites was and is built in the Soviet Union, East Germany and Czechoslovakia. Observation posts are to be found in all seven countries. Up to now the objective of this Interkosmos series has been the study of solar radiation, the earth's atmosphere, and cosmic radiation. The information obtained is assimilated jointly by scientists of the countries concerned. Launchings of Interkosmos satellites (using Russian carrier rockets of the Kosmos type) are attended by scientific representatives of all seven nations. Experiments on a more modest scale also take place, for instance the flight of Vertikal-1, a geophysical rocket launched on November 28, 1970, which reached a height of 487 km. After taking measurements of the sun's ultraviolet and x-ray radiation for a short time, the instrument capsule made a successful parachute landing. It is intended that co-operation between the socialist countries will be extended in the near future.

Relatively close contact also exists between French and Russian space experts. This contact, too, has been officially ratified in the form of agreements between the commission chaired by Professor Petrov and the French National Centre for Space Exploration (CNES). The first agreement dates from June 30, 1966. Meanwhile a Franco-Russian satellite observation post has been established on the island of Kerguelen in the southern Indian Ocean. Joint experiments are carried out in the field of meteorology and with the Molniya communications satellites which maintain regular contact between Moscow and Paris; the French instruments are to fly in Russian satellites. Naturally all information obtained is exchanged. A significant result of Franco-Russian co-operation is the fact that Lunokhod-1 and-2 are equipped with a French laser-reflector. The reflector is used by both Russian and French scientists in determining the exact distance between the earth and the moon.

In addition, French artificial satellites have been and are being launched by means of Russian rockets. On December 27, 1971, the "Auréole" was launched, intended for research into the upper layers of the atmosphere. On April 4, 1972, the SRET (which the Russians call MAS—*Malii Avtonomni Sputnik*, or "small, independent satellite") was given a lift by the umpteenth Molniya communications satellite. The purpose of this little French artificial moon was to test various types of solar cells.

Standardised docking gear

Shortly after Neil Armstrong and Edwin Aldrin had become the first to set foot on the moon, some headway was made in "co-operation in space" between the Soviet Union and the United States. Earlier, in the first week of July 1969, Frank Borman had visited his colleagues in Star Village, where he had raised the question of co-operation unofficially. His visit resulted in an invitation to cosmonauts Feokistov and Beregovoi, who visited the US in October 1969 and had a conversation with President Nixon. Soon afterwards, in December 1969, Professor Keldysh wrote in a letter to Thomas Paine (who at that time was director of NASA) that he agreed that up to now co-operation in space had not amounted to much, and he remarked that there was a "need for further development of this co-operation".

From April 1970 onwards, Dr. Paine wrote a number of letters to Keldysh, in which he invited the Russians to propose experiments which could be assimilated in future American projects. In one of his letters Dr. Paine stated that the Russians were of course free to make use of the laser-reflector left on the moon by the crew of Apollo-11. If they wished, they could also have moon samples for examination in their own laboratories. The Russians were, moreover, invited to attend meetings at which American plans for the exploration of Mars within the Viking project were to be discussed.

In April 1970 Dr. Paine had an informal meeting with Anatoli Blagonravov, Member of the Academy and Chairman of the Commission for the Exploration and Exploitation of Cosmic Space of the Academy of Sciences of the USSR. In the course of their conversation Dr. Paine suggested that the possibility should be discussed of standardising the docking equipment of Russian and American spaceships, in order to make mutual rescue operations in space possible. The same subject was subsequently repeatedly raised during meetings between Russian and American experts, in which, among others, Dr. George Low (acting Administrator of NASA after Dr. Paine's departure) and Frank Borman were involved. Finally Russian and American experts got down to essentials at a meeting in Moscow in October 1970. For the first time concrete plans were drawn up for standardised docking equipment, plans which were further elaborated in the course of 1971. In January 1971 there were further talks in Moscow, at which agreement was reached between the USA and the USSR on other space activities, such as moon exploration.

A joint method for rescuing space travellers in an emergency is certainly advisable now in view of the fact that traffic in space will increase more and more once permanent space stations have been established.

The foundations for such a system were laid during talks in Moscow between Russian and American experts on October 26, 27 and 28, 1970. The Russian delegation consisted of Professor Boris Petrov, Cosmonaut Konstantin Feokistov (deputy director of the Soviet Bureau for Manned Space Travel) and three specialists in the field of rendezvous and docking systems: V. S. Siromiatnikov, V. V. Suslennikov and V. A. Lavrov. The American delegation was led by Robert R. Gilruth (director of NASA's Manned Spacecraft Centre), Arnold W. Frutkin (NASA international affairs) and technicians George B. Hardy, Caldwell C. Johnson and Glynn S. Lunney.

Considering what had gone before, one may say that the discussions took place in a surprisingly frank atmosphere. Professor Petrov enabled Dr. Gilruth and his team to visit Star Village to make the acquaintance of cosmonauts and technicians. The Americans were allowed to sit in the Russian Soyuz spaceships and the functions of all systems were explained to them in detail. On the first day of talks the two delegations exchanged very detailed technical information and designs. The American material related to the Gemini and Apollo rendezvous and docking systems, and the Russian material concerned the comparable systems of the Soyuz. The Americans examined not only the well-known Soyuz system without a transfer tunnel (such as used in Soyuz-4 and -5, in which the cosmonauts had to transfer from one capsule to another from the outside), but also a new docking system in which it would be possible to move from one to the other inside. This system was used for the first time in the Salyut programme.

Further discussions emphasised the necessity of adapting standardised docking equipment to future spacecraft and space stations. It will, of course, be essential to test such equipment in experimental flights. Initially the possibility of linking an Apollo to a Russian Salyut was considered. But the Salyut has only one docking hatch, to which the Soyuz is connected in flight and to link up an Apollo as well would have presented numerous constructional problems. It was therefore decided to aim at a simple Apollo-Soyuz flight, planned to take place on July 15, 1975. A special transfer tunnel (docking module) is being developed for this purpose; this will be launched with Apollo. In the course of their joint flight the cosmonauts will enter each others' spacecraft and they will carry out joint experiments. Apollo will have a three-man crew; Soyuz, which is smaller, will be manned by two cosmonauts. It is possible that the joint Apollo-Soyuz flight will be followed by further Russo-American co-operation in the field of manned space-flights. The new docking apparatus will not only enable Russian and American spacecraft to link up; it will also be possible for a Russian craft to dock into an American space station and vice versa. It will thus be possible to carry out joint projects as well as rescue operations.

Space travel: international

Advances are also being made in the field of unmanned space exploration. In January 1971, after discussions in Moscow, Professor Keldysh and George Low issued a joint communiqué which stated, among other things, that NASA and the Academy of Sciences of the USSR would co-operate in:

—a study of future subjects for research and (subsequently) their results;
—improved exchange of meteorological information;
—a joint study of the natural environment with the use of satellites and conventional methods;
—more extensive information concerning cosmic biology;
—the examination of lunar material; to begin with, the Russians were to receive three grams of both the Apollo-11 and the -12 samples, and the Americans three grams of the Luna-16 material.

Konstantin Tsiolkovsky's book *Vnye Zyemli* (Outside the Earth) starts as follows:

"A magnificent castle is hidden somewhere among the high Himalayas. Here live a Frenchman, an Englishman, a German, an American and a Russian. Disillusioned with man and with the pleasures of life, they have been driven together into seclusion. They have found their happiness in science." This is the international company which, in Tsiolkovsky's book, forms the crew of the first space station.

We have not yet reached that point, but there is every sign that the "father of space travel" will be proved right in this respect also. "The problems which we encounter are problems for all mankind," said Cosmonaut Shatalov, the commander of the first space station. And in the words of American astronaut John Young, "I think Shatalov hit the nail on the head. I believe that one day we shall reach a point where each nation—however small—will be able to play its own part in this field."

This will be necessary if ever we are to achieve space travel on a large scale. For the challenge facing us is, in principle, as great as space itself.

INDEX

Solomon R. Baker Library
3 1725 00107 6772

TL789.8
R9S613
1974